Safeguarding the Organization against Violence and Bullying

Also by Paul McCarthy

SOCIALLY RESPONSIBLE MANAGEMENT AND ETHICAL INVESTMENT: Employment Practices (*with M. Henderson, M. Barker, and M. Sheehan*)

BULLYING: From Backyard to Boardroom (*Second Edition with J. Rylance, L. Bennett, and H. Zimmermann as editors*)

BULLYING: Causes, Costs and Cures (*with M. Sheehan, S. Wilkie, and W. Wilkie as editors*)

POSTMODERN DESIRE: Learning from India

Also by Claire Mayhew

PREVENTING CLIENT-INITIATED VIOLENCE: A Practical Handbook

PREVENTING VIOLENCE WITHIN ORGANISATIONS: A Practical Handbook

VIOLENCE IN THE WORKPLACE – PREVENTING COMMERCIAL ARMED ROBBERY: A Practical Handbook

OCCUPATIONAL HEALTH AND SAFETY IN AUSTRALIA: Industry, Public Sector and Small Business (*with C. Peterson as editors*)

Safeguarding the Organization against Violence and Bullying

An International Perspective

Paul McCarthy and Claire Mayhew

First published 2004 by
PALGRAVE MACMILLAN
Houndmills, Basingstoke, Hampshire RG21 6XS and
175 Fifth Avenue, New York, N. Y. 10010
Companies and representatives throughout the world

PALGRAVE MACMILLAN is the global academic imprint of the Palgrave Macmillan division of St. Martin's Press, LLC and of Palgrave Macmillan Ltd. Macmillan® is a registered trademark in the United States, United Kingdom and other countries. Palgrave is a registered trademark in the European Union and other countries.

ISBN 1–4039–3252–2

This book is printed on paper suitable for recycling and made from fully managed and sustained forest sources.

A catalogue record for this book is available from the British Library.

Library of Congress Cataloging-in-Publication Data
McCarthy, Paul, 1943–
    Safeguarding the organization against violence and bullying : an international perspective / Paul McCarthy and Claire Mayhew.
        p. cm.
    Includes bibliographical references and index.
    ISBN 1–4039–3252–2 (cloth)
        1. Violence in the workplace. 2. Bullying in the workplace.
3. Industries–Security measures. I. Mayhew, Claire. II. Title.

HF5549.5.E43M387 2004
658.4'73–dc22   2004042846

10   9   8   7   6   5   4   3   2   1
13   12   11   10   09   08   07   06   05   04
Printed and bound in Great Britain by
Antony Rowe Ltd, Chippenham and Eastbourne

Contents

List of Tables

List of Figures

Preface

Paul McCarthy and Claire Mayhew

Internationally, there has been a groundswell of interest in unreasonable work practices, bullying at work and occupational violence more generally (Di Martino *et al*, 2003; McCarthy, 2003a; Einarsen *et al*, 2003; Mayhew, 2002a; Chappell and Di Martino, 2000). It has been estimated that one in four employees are likely to encounter repeated bullying at some time in their working lives (Leymann, 1997), and 6% physical violence each year. A wide circle of witnesses, work colleagues, family members, and friends, as well as supervisors, managers, health and safety and harassment officers, counsellors, medical doctors, insurers, lawyers and regulators can also be drawn into interactions with recipients and perpetrators of bullying and violence. The rise of service industry work in the new global economy has also placed employees, clients, contractors, and suppliers in complex interactions in which work and life pressures, skills limits and poor environmental design intensify risks of aggressive behaviours.

We have aligned occupational violence and bullying on a *continuum of severity* spanning: homicide, assaults, threats, verbal abuse and less overt bullying (e.g. denigration, social exclusion, and unreasonable work practices). Our experiences in research, policy development, training and advocacy for prevention of work-related bullying and violence in the last decade shape our approach to safeguarding. Paul McCarthy applies a critical research methodology in mapping the evolution and enfolding of meanings of 'bullying at work' and 'occupational violence' (Hatcher and McCarthy, 2000). Patterns through which the likelihood, severity, and costs of incidents escalate from the interplay of individual, organizational and societal risk factors are identified. A cross-sectoral web of safeguards that are sustainable due to their contribution to social, economic and environmental performance is considered necessary to constrain the escalation of risk and severity. The chapters contributed by Claire Mayhew are grounded in her recent empirical research studies in the health, detention, housing, retail, transport, and tertiary education industries. Her study findings inform comprehensive recommendations for risk management that emphasize administrative controls and environmental design. Our approaches meet in the identification of risk factors that are common for both bullying and violence, the realization that each can be a risk factor for the other, and opportunities for safeguarding synergies. From this common ground, we advocate *safeguarding* that casts a web of mutually supportive preventions across the harder and softer edges of violence and bullying in contemporary work environments.

Our use of the notion of 'safeguarding' reflects our interest in connecting with feelings of threat and insecurity that pervade contemporary work and social life in ways that both echo and contribute to risks of bullying and violence. Threats to job position, consumption opportunities, social status, health, security and political futures resonate across workplaces and communities to raise temperatures in an *age of fear and rage* of global extent. Thus, employees, clients, contractors, suppliers and others engaged in exchanges with organizations are more likely to be drawn into situations in which they can be recipients and/or perpetrators of bullying and violence. It is unlikely that the quantum of bullying and violence has increased in the long history of organizations, although the composite of forms, impacts and responses has altered. With the sanitization of violence (Foucault, 1977) there has been an accentuation of more subtle and systemic forms of bullying. At the same time the incidence and severity of new overt forms of violence in work environments has increased. Heightened threats of randomized rages, opportunity crimes, destructive swarming, internet aggression, and global cellular terrorism are evident in organizational operating environments.

The rising tide of interest in workplace aggression has spawned publications and a vocabulary more often applied in naming hitherto normalized negative actions as 'bullying' and 'violence'. These sentiments have fuelled broad-based advocacy for prevention. Governments in most western countries have responded by releasing anti-bullying/violence guidelines oriented to meeting Occupational Safety and Health (OSH) obligations. The case for the implementation of preventive measures is commonly advocated in terms of legal risks, cost reduction, productive benefits, and enhancement of customer service and corporate image. Employee representatives have more often placed requirements for prevention on the table in enterprise bargaining negotiations. This flourishing of concerns has also prompted research and an emergent international scientific literature addressing workplace bullying and violence. The commissioning of a series of international research studies and discussion papers has also ensued.

Definitional terms for workplace bullying and violence are the subject of much discussion in the international arena. Researchers from several countries have recently collaborated to produce a comprehensive statement of research and practice concerning bullying and emotional abuse in the workplace (Einarsen et al, 2003a). It defines bullying as:

> ...[H]arassing, offending, socially excluding someone or negatively affecting someone's work tasks. In order for the label bullying (or mobbing) to be applied to a particular activity, interaction or process it has to occur repeatedly and regularly (e.g. weekly) and over a period of time (e.g. about 6 months). Bullying is an escalating process in the course of which the person confronted ends up in an inferior position and becomes the target of negative social acts. A conflict cannot be called bullying if the incident is an isolated event or if two parties of approximately equal 'strength' are in conflict.   (Einarsen et al, 2003b: 15)

Researchers concerned with more overt occupational violence have also collaborated in the development of definitional and policy terms within the auspices of the United Nations' International Labour Organisation (ILO). A Code of Practice by the International Labour Organisation (ILO), is also likely to be finalized in 2004. The Code of Practice states that workplace violence is:

> Any action, incident or behaviour that departs from reasonable conduct in which a person is assaulted, threatened, harmed, injured in the course of, or as a direct result of, his or her work.
>
> • Internal workplace violence is that which takes place between workers, including managers and supervisors.
> • External workplace violence is that which takes place between workers (and managers and supervisors) and any other person present at the workplace.   (ILO, 2003a: 4)

The span of definitional terms for bullying and violence indicates that incidents can involve both subtle forms of aggression (e.g. 'bullying') as well as more explicit acts of occupational 'violence' (e.g. assaults). Particularly for less overt forms of aggression, interpretations and imputed meanings can vary between individuals because of subtle variations in perpetrator behaviours, the degree of threat to personal self-esteem or livelihood, and acculturation to 'normalised' forms of behaviours at different worksites. Hence, a range of inappropriate behaviours makes up the broad gamut of violence and bullying at work.

Differences in the definitions of bullying and violence are indicative of difficulties in demarking the gamut of more and less overt behaviours and work practices as 'bullying' or violence'. These difficulties stem from different popular and disciplinary meanings that have been enfolded in the terms. The legacy of concerns with risks, incidents, and impacts that are more tangible or physical in the discourse of OSH foregrounds the notion of 'violence' in the ILO definition. Whereas, the definition of bullying by Einarsen *et al* (2003) casts bullying (including mobbing) as an essentially psychological phenomenon in which traumatic impacts arise from the experience of repeated (often minor) incidents and their reliving as 'though terror' over time (Leymann, 1990). A considerable overlap can be discerned in definition and research practice concerning occupational violence and bullying. Studies of bullying commonly account for incidents of more overt violence, and studies of occupational violence also record negative acts that are less overt and more psychological. The potential for one-off incidents and threats thereof that can be of considerable psychological and physical magnitude point to limitations in the definition of bullying. On the other hand, the ILO definition of violence has little accounting for the repetitive nature of bullying and the manifestation of its traumatic psychosomatic impacts over long periods of time. Concerns as to whether bullying can be dealt with adequately within the rubric of 'stress' can also be raised. For example, 'bullying at work' has been depicted as hybridization of prior understandings of stress, abuse, harassment and school bullying that expresses a broad-based politics of resentment about negative experiences in work environments impacted by global market pressures (McCarthy, 2003a). However, common recognition of the severity of impacts of occupational violence and bullying in terms of post-traumatic stress and evidence that bullying and violence can each be risk factors for the other do provide grounds for convergence in understandings of the phenomena.

The evidence appears to be overwhelming: overt occupational violence is most common in jobs: (a) where cash is on at hand, and (b) which require substantial face-to-face contact between workers and their clients or customers (Mayhew, 2002a; Chappell and Di Martino, 2000). The incidence and severity of violence varies markedly between these jobs and those where workers have little contact with outside people and where money does not change hands.

Less overt forms of bullying, including unreasonable work practices have become an issue as global market pressures and deregulation have prompted ongoing restructuring, downsizing, outsourcing, and per-formance management in workplaces. People skills have been sorely tested and thresholds at which employees, clients and other stake-holders might reflex into aggressive behaviours lowered. Experiences in workplaces have resonated with wider socio-economic and political threats. In these circumstances, 'bullying at work' has emerged as a new signifier of distress that has acted as a solar collector of resent-ments (McCarthy, 2003a). Survey researchers commonly report widely varying prevalences of bullying at work, since the nature, number, duration and impacts of incidents used in determining whether a re-cipient has 'experienced bullying', and the extent of randomness amongst respondents, vary widely.

There are gender variations: females tend to experience higher levels of verbal and sexual abuse, and males tend to receive more overt threats and physical assaults (Mayhew, 2002a; Chappell and Di Martino, 2000; Mayhew and Quinlan, 1999). This variation in risk can be partially explained by the gender division of labour, with women concentrated in lower status and 'caring' jobs with greater face-to-face contact with clients/customers/patients. For example, female respondents to studies of bullying commonly report ex-periencing higher incidences of bullying than do males. Rising reports of bullying have also been associated with increasing entry of females into hitherto male-dominated workplaces. The greater capa-city of females to speak out about distressing experiences in contem-porary workplaces has also been noted. Civil service organizations have been at the forefront of the development of policies to address equity and sexual harassment that have encouraged the entry of females to workplaces. This lineage of policy development and enact-ment has also provided the basis for understanding and naming 'bul-lying'. Thus, it is not surprising that reports of bullying have been higher in the civil service than in the private sector. (McCarthy, 2001a; McCarthy and Mayhew, 2003).

Off-site and isolated work environments are associated with higher risk of overt occupational violence. One United Kingdom study of 1,000 workers found that 1 in 3 who went out to meet their clients had been threatened, and 1 in 7 assaulted at some stage of their working life (Bibby, 1995). 'Off-site' workers who have been victimized include: a British lawyer (Vandenbos and Bulatao, 1996), Australian taxi drivers (Mayhew, 2000a), US community service workers (CAL/OSHA, 1998), health workers in car parks (NOHSC, 1999), and those working alone in high crime areas (OSHA, 1998).

Thus the incidence and severity of occupational violence/bullying varies across jobs because the risk factors differ. Patterns of violence/bullying may also vary because some organizations apply better-targeted intervention strategies. Only work-related homicides are reliably reported. At best, between 10 to 20% of incidents are formally recorded, and as a result the official databases significantly under-state the extent of occupational violence (Mayhew, 2002a). There are also overlapping jurisdictional responsibilities between the criminal justice system, the OHS authorities, and individual organizations – all of which record some data on occupational violence and bullying in different ways. Thus, while the general risk factors and overall patterns of occupational violence and bullying are known, the data are poor.

The ILO has also identified organizations at increased risk as those that:

> ...[P]rovide services in direct contact with members of the public, provide services early in the morning or late at night, are located in high crime areas, operate from relatively unsecured premises, are small and isolated, are understaffed, operate under the strain of reform and downsizing, work with insufficient resources, function in a culture of tolerance or acceptance of stress and violence, have a poor working climate, are management based on intimidation and generation of stress, have a climate of discrimination, including gender and racial discrimination, have smoking or drinking prohibition/limitation; or do not have such rules, have drugs, alcohol or/and weapons available onsite, possess unclear codes of conduct, have inconsistent rule enforcement. (ILO, 2003b: 18)

Chappell and Di Martino (2000) profile the range of violent behaviours that can be experienced at work in their comprehensive book *Violence at Work*, identify how patterns are changing over time, and indicate some potential solutions. It is important to note that the *severity* of violence is not the same as the *incidence*. Some jobs have a high incidence of lower-level violence, for example, verbal abuse of workers in fast-food outlets, although such incidents rarely have fatal outcomes. Conversely, armed hold-ups of taxi drivers may be less common, but the potential for a fatality is very much higher. This is not to trivialize any form of occupational violence; all are an affront to human rights and dignity. While assaults on-the-job are as yet relatively rare events in Australia and Britain, the incidence may increase as risk factors alter.

It is also known that the traumatic impacts of repeated bullying can accumulate over time to rival those of overt acts of violence (Leymann and Gustafsson, 1996). The conceptualization of bullying in terms of cumulative psychosomatic effects of a litany of often-minor incidents that are difficult to evidence has meant that researchers into mobbing and bullying have been concerned with the psychological character- istics of recipients. Personal, work and life histories of recipients and ways their experiences of negative actions have mediated their trans- ition from 'recipient' to 'victim' over time have been of particular interest. For example, Einarsen and Mikkelsen (2003: 133–4) have pro- vided evidence of relations between victim personality and bullying in the following terms. Provocative behaviours can lead to conflicts that escalate into bullying. The display of personality traits that arouse antagonism in certain situations may also trigger bullying. A recipi- ent's self-esteem, social anxiety, negative affectivity, self-efficacy, per- ceived locus of control, attributional style, and coping strategies also moderated the severity of health impacts of bullying. Matthiesen and Einarsen (2001) also found that victims of bullying had elevated per- sonality profiles on the MMPI2 that suggested a vulnerability factor for personality clusters they termed 'seriously affected' and 'disap- pointed and depressed' as distinct from a 'common group' cluster that had a 'normal' personality profile in spite of the largest exposure to negative actions. However, the extent to which the traits of victims might be symptomatic of the destruction of the victim's personality in the experience of bullying, or attributable to prior vulnerability, remains the subject of ongoing debate (Leymann and Gustafsson, 1996).

On the other hand, occupational violence researchers have been more often concerned with tangible evidential risks and causes for which responsibility can be assigned and administrative controls, envi- ronmental designs and sanctions against perpetrators implemented. However, the causes and responsibilities for subtle bullying can be 'in the eye of the beholder', thus less amenable to sanctions, and more to interventions that reconcile differences and build skills, coping and resilience. Flexibility in responses is thus necessary to address more and less overt forms of bullying and violence and their interactions com- prehensively. A balance of safeguards at the individual, organizational, and regulatory levels that attend to coping and resilience, administra- tive controls and environmental design is more likely to be effective in preventing the multiplicity of forms and directions of bullying and violence.

Arguably, a core reason why occupational violence/bullying has not been widely investigated by occupational safety and health (OSH) or management researchers is that much of the scientific literature has remained divided between those working within a criminal justice system paradigm, those in OSH, researchers studying bullying, and the specific literature in the health industry. The scant attention to bullying and violence in the corporate social responsibility and governance literature should also be noted. For the most part, these are distinct disciplines with diverse research and literature sources. One of the challenges is to bring these distinct bodies of knowledge together in a way that enhances our capacity to reduce the risks in a range of situations and job tasks where occupational violence and bullying can flourish. The value that safeguarding against bullying/violence can add to corporate sustainability also needs to be demonstrated to move beyond minimum compliance and motivate the implementation of safeguards.

Another core problem for researchers is that non-reporting of non-fatal incidents is very common. The international evidence suggests that, at best, around 10% of incidents will be formally reported (see Chappell and Di Martino, 2000). Those that remain un-reported are commonly known as the 'dark figure' (Mayhew, 2002a). In the case of bullying, reporting of incidents can be confounded by difficulties in defining and evidencing less overt behaviours and work practices that are 'unreasonable', and differentiating them from a 'normal' range of conflicts expected in contemporary workplaces. Those who do report experiences of bullying place themselves at risk of stigmatization, added duress and costs in seeking remedies, and can be disadvantaged where classified as whistleblowers (McCarthy, 2003b). Since the incidence and severity of occupational violence/bullying is largely hidden and definitional terms vary, inappropriate ways of dealing with incidents can grow up. As a result, risks and severity of incidents can escalate, large amounts of money paid out to permanently disabled workers, extensive insurance claims made before the need for remedial interventions is recognized, and reputations ruined. Arguably it is only when the risk factors have been fully identified that site-specific control and value-adding strategies can be designed and implemented in a sustainable manner.

In this book we aim to bring together the extant research evidence on occupational violence and bullying, estimate the costs to organizations and the community at large, examine the overlaps between manifestations of aggression in the broader community and the extent of spillover into workplaces, and identify preventive interventions that may *safeguard* organizations from these threats. Opportunities for safeguarding strategies to add value to organizational skills, productivity, quality and reputation amongst clients, investors and governmental stakeholders are emphasized. Finally, we call for a broad-based global interdisciplinary initiative to develop definitional and policy terms that adequately account for commonalities and differences in occupational violence and bullying. Such an initiative is urgently needed to orientate organizational safeguarding strategies and domestic regulatory responses to international best practice.

1
The Safeguarding Challenge

Paul McCarthy

Evidence of the impacts of occupational violence and bullying on pro-ductivity, safety, health, reputation and market performance has led to increasing interest in safeguarding by management, employees, profes-sional service providers and governments in recent years. An estimated 6% of European employees experience physical violence per year (2% from fellow workers, and 4% from 'outsiders') and at least 10% are sub-jected to bullying (Hoel *et al*, 2001). Costs have been estimated in excess of BP£1.17 million per 1,000 employees per annum (Rayner, 2000). Violent attacks by disgruntled employees, clients, or other out-siders can also threaten the viability of an enterprise.

While there are demonstrable reasons for the implementation of preventive policies, manifest difficulties make safeguarding a for-midable challenge. As understandings of occupational violence and bullying continue to evolve, definitions vary and policy responses remain fragmented and unevenly implemented across industry sectors. Inconsistent environmental design and administrative inter-ventions are also evident. Insufficient comparable scientific evidence from random-sample survey studies across industry sectors also means there is an insecure basis for shared understandings and common approaches to safeguarding. In this climate, employers and employees remain exposed to unacceptably high risks of bullying and violence. Nor are preventive interventions sufficiently refined to realize the positive productive, market, financial, and governmental benefits of safeguarding.

The purpose of this book is to advance pathways to safeguarding. However, prevention of occupational violence/bullying requires shared understandings and mutual benefits for employees, management and governmental interests. The following questions prescribe key vectors of discussion around which the logic of safeguarding is woven in the ensuing chapters.

- How have meanings of workplace bullying and occupational violence evolved and interpenetrated?
- What are the risk factors for occupational violence and bullying, and how do they relate to the risk factors for terrorism?
- On what basis are the costs of occupational violence and bullying assessable in economic, as well as social and political terms?
- What does recent research in industry sectors tell us about the incidence and severity of occupational violence and bullying, and what implications can be drawn for safeguarding?
- How can a road map to safeguarding be formulated to the mutual benefit of employee, managerial, industry and governmental interests?

The rise of safeguarding as a managerial issue

Organizations now more often implement measures to prevent occupational violence and bullying to meet a variety of productive, health, safety, financial, reputational, governance, social responsibility, and legal concerns. 'Horror' stories about overt violence and less overt bullying have been commonly circulated in the media. Research findings also point to an increasing risk of incidents with damaging costs to individuals, organizations and the wider society (see: Chapters 2 and 3). Media stories and research findings have resonated widely with experiences of victims and service providers in recent years, and contributed to a ground swell of concern that has driven the development of preventive strategies.

Responses to victim's needs and advocacy for remedies is now more often forthcoming from victim support groups, professional service providers, employee representatives, human resource managers, employer associations, and governmental agencies. Employee assistance providers, counsellors, psychiatrists, medical officers, lawyers, educators, trainers, criminologists, mediators, researchers, and ethicists commonly provide professional services following bullying and violence. This service provision has prompted and legitimated governmental preventive responses. Consequently, departments of occupational health and safety, workers' compensation, industrial relations, anti-discrimination, and crime and misconduct, as well as ombudsmen, have often promoted preventive interventions.

Government interest in prevention of occupational violence and bullying has more often been led by Occupational Safety and Health (OSH) agencies. While governmental responses have mostly been self-regulative, advisory standards and codes of practice with legal force have emerged in some jurisdictions. In many cases, OSH agencies have involved professional service providers, employee and industry representatives, support groups, and other government agencies in consultative processes to develop these responses. Estimation of the considerable costs of workplace bullying and violence to victims, the organization and the wider society (see: Chapter 3), has justified the resourcing of prevention by government agencies such as the Health and Safety Executive in Britain.

Danger zones

While occupational violence and bullying are now recognized as risk management issues, safeguarding initiatives remain severely compromised due to:

- Lack of an international standard on occupational violence or workplace bullying to which there is widespread subscription by national employee, industry, OSH agencies, and professional bodies;
- Difficulties in evidencing bullying due to its psychological and perceptual nature and the tendency for traumatic effects to accumulate over time from the repetition of (often minor) incidents;
- An emphasis on overt forms of occupational violence that leave less overt and systemic forms of bullying unaccounted;
- The need to account for multiple forms and directions of bullying and violence, that originate within and without organizations and traverse their boundaries;
- The emergence of terrorism as a threat to workplaces, in part due to blow back effects of the global circulation of occupational violence and bullying;
- Estimates that only one in five incidents of occupational violence are likely to be reported (Chappell and Di Martino, 2000);
- A dearth of scientific research evidence using comparable constructs and random sample survey approaches across industry sectors;
- The predominant implementation of preventive measures by government authorities, with tokenistic responses in the private sector, and little by small business;

- Limited main-streaming of prevention in everyday management practice, for example, in recruitment and selection, performance appraisal, accounting, and evaluation procedures;
- Tensions between developmental and restorative remedies, and those focussed on punitive sanctioning of perpetrators; and,
- Normalization of bullying and violence in work environments subjected to intense pressures of restructuring and downsizing.

These challenges and the disconnected nature of responsibility across sectors result in employees, organizations and communities remaining at unacceptable risk of exposure to occupational violence and bullying.

The evolution of meanings

The discussion of meanings of occupational violence and bullying that follows serves to introduce the conceptual frameworks underpinning arguments advanced in this book. Particular attention is given to definitions that address both occupational violence and bullying.

Definitions

The International Labour Organisation (ILO), the World Health Authority (WHO), and the Health and Safety Authority (HSA) in Ireland and have promulgated definitions that address both workplace bullying and violence. These definitions are salient conceptual genotypes from an evolution of meanings of bullying and violence in the last decade.

The ILO's code of practice depicts bullying as 'ongoing violent abuse' within a generalized notion of 'workplace violence'. Notably, this definition carries forward and extends the typology of internal, external, and client-initiated violence promulgated by California Division of Occupational Health and Administration CAL/OHSA (1998) (see: Chapter 2). Workplace violence is defined as:

Any action, incident or behaviour that departs from reasonable conduct in which a person is assaulted, threatened, harmed, injured in the course of, or as a direct result of, his or her work.

- Internal workplace violence is that which takes place between workers, including managers and supervisors.
- External workplace violence is that which takes place between workers (and managers and supervisors) and any other person present at the workplace.   (ILO, 2003a: 4).

The definition promulgated by the WHO addresses bullying within 'occupational violence':

> The intentional use of power, threatened or actual, against another person or against a group, in work-related circumstances, that either results in or has a high degree of likelihood of resulting in injury, death, psychological harm, maldevelopment, or deprivation (WHO, 1995).

The Code of Practice of the Irish HSA includes overt forms of violence within the ambit of 'workplace bullying':

> Workplace bullying is repeated inappropriate behaviour, direct or indirect, whether verbal, physical or otherwise, conducted by one or more persons against another or others, at the place of work and/or in the course of employment, which could reasonably be regarded as undermining the individual's right to dignity at work. An isolated incident of the behaviour described in this definition may be an affront to dignity at work but as a once off incident is not considered to be bullying (Health and Safety Authority, 2002: 4).

These definitions should be seen as convergent formations of meaning, arising from the intersection of different disciplinary and sectoral interests. As such, the definitions represent the outcome of gravitational pushes and pulls in the force field of knowledge and interest.

The interpenetration of meanings

In the most recent decade, the sharing of concerns about aggressive behaviours stimulated a cross-fertilization between the discourses of workplace bullying and occupational violence. The unfolding of incipient meanings of workplace bullying and violence within OSH, sexual harassment, and criminology enabled the recognition of bullying as a 'softer' psychological violence, or harassment of a non-sexual nature. Bullying has also been associated with violence and other forms of criminality. A further loop in the interpenetration of meanings occurred with realization that workplace bullying and violence can each be risk factors for the other.

Disciplinary cross-fertilization

The roots of occupational violence stem from concerns about workers' health in industrial sociology, occupational medicine, duty of care, workers' rights, and criminology that intersect in the discourse of OSH. These roots of meaning figure in the emphasis on risk management couched in duty of care terms in the discourse of occupational violence.

On the other hand, 'bullying at work', arose as a new signifier of distress that hybridized prior meanings of school bullying, domestic violence, sexual harassment, and discrimination (McCarthy, 2003a). The term 'bullying' has been most evident in use in the UK, Norway and Australia, and has had a strong resonance with the conceptualization of 'mobbing' (or ganging up) as has been the predominant terminology in Sweden and Germany (Leymann, 1990). Notably, 'bullying at work' has gained currency during a time of restructuring, downsizing, deregulation, contract and part-time labour, widespread entry of women into workplaces, and interest in the management of emotions at work.

Service provision by psychologists, psychiatrists and lawyers, coupled with governmental interest in providing regulations protecting employees' health and safety have prompted growing recognition of psychological injury due to stress as a basis for workers' compensation claims. This convergence of interests has also stimulated the broadening of definitional terms for occupational violence to include more subtle and systemic types of bullying. In turn, regulatory obligations for psychological injury, such as OSH duty of care, have more frequently been recognized within the discourse of bullying.

Occupational violence and bullying in OSH

The discourse of occupational violence has traditionally emphasized overt physical violence and threats from *external* sources (e.g. hold-ups) as well as *client-initiated violence*, while 'bullying' first entered this discourse as *internal violence* (CAL/OSHA, 1998) (see: Chapter 2). This categorization of bullying as a form of occupational violence was enabled by growth of awareness in the 1990s that escalated, destructive interpersonal conflict (i.e. bullying or mobbing) caused psychological injury and constituted *psychological violence*. The recognition of bullying as *internal violence* also reflected initial understandings of bullying and mobbing as dysfunctions *within* organizational life (Swedish National Board of Occupational Safety and Health, 1993; Zapf and Gross, 2001).

As pressures in the workplace grew in the last decade, so did the volume of writing about the dark side of organizational life. A sampling of topics includes: *aggression* (Björkqvist *et al*, 1994); *petty tyranny* (Ashforth, 1994); *incivility* (Andersson and Pearson, 1999); *rage* (Robinson,

1999); *hate* (Cunneen *et al*, 1997); *men and women arguing* (Mapstone, 1998); *territorial wars* (Simmons, 1998); *toxic work* (Reinhold, 1997); *cultures of secrecy and abuse* (Beed, 1998); *brutal bosses* (Hornstein, 1996); *work abuse* (Wyatt and Hare, 1997); *scapegoats* (Daniel, 1998); *whistle-blowing* (De Maria; 1999); and *stupid white men* (Moore, 2001).

Publications such as these, and accompanying media discussions, have contributed to an appreciation that workplace bullying and violence are multi-dimensional and interactive phenomena that can originate within and without organizational boundaries. For example, external parties, including clients, have been recognized as perpetrators of bullying *towards* employees. Stressed employees have also been known to inflict bullying and violence on clients, members of the public, and other external parties. Systemic factors have also been recognized as sources of bullying and violence (Bowie, 2002). This proliferation of forms and directions tends to extend categories in the CAL/OSHA (1998) typology (outlined in Chapter 2 in more depth).

Resonance with sexual harassment

Women who now more often enter male dominated workplaces have reported experiencing harassment that was neither overtly physical nor immediately sexual. A strong resonance between notions of 'harassment' and workplace bullying has ensued. Women's inclination to speak out about emotional distress due to inappropriate behaviours and their positioning down organizational hierarchies has contributed to their more often acknowledging experiences of bullying than males (Task Force on the Prevention of Workplace Bullying, 2001; McCarthy and Mayhew, 2003). Experience in networking resentments to domestic violence, sexual harassment, discrimination and stalking has also placed women at the forefront of advocacy to prevent workplace bullying. Sexual harassment and equity policies have also provided useful templates for anti-workplace bullying policies. (McCarthy, 2003a). Indeed, in some jurisdictions bullying has been termed 'workplace harassment' (Department of Industrial Relations, Queensland, 2003).

Women's experience of domestic violence and sexual harassment was also consonant with the notion of 'emotional abuse' at work (Nicarthy *et al*, 1993). Sexual harassment has also been recognized as a form of occupational violence (Chappell and Di Martino, 2000). In addition, discrimination experienced by women has resonated with the concept of systemic violence. Patriarchy has also been factored into explanations of the greater experience of bullying and violence by women in male dominated workplaces (Amnesty International, 2003).

However, a detailed discussion of the relations of patriarchy, sexual harassment, and occupational violence/bullying is not undertaken in this book, and is left to specialist researchers elsewhere.

Criminality

There are close overlaps between occupational violence and criminal behaviour. A recent study of adolescent detainees found that some 50% of the respondents nominated themselves as school bullies, 19% indicated they had been victims of school bullying, and 12% reported being both victims and bullies (Kenny, 2003). Bullying has also been identified as a key factor in the cycle of violence within prisons (Tattum and Herdman, 1995).

While overt violence and physical threats fall within criminal laws, bullying does not. Definitional and evidential difficulties for bullying have also impeded the substantiation of allegations in criminal and civil courts. On the other hand, the causes and responsibility for violence are not always easily decidable (Cately and Jones, 2001). There are also questions as to what might be a reasonable response to an unreasonable provocation or threat. The tendency to focus on behaviours and responsibilities of perpetrators and the punishment of illegal behaviours, mostly leaves deep-seated systemic causes outside the preventive loop. Difficulty in assigning responsibility, limited concern with systemic causes, and the focus on punishment have contributed to the call for restorative and transformative remedies oriented to development and reconciliation (Ahmed, Harris, Braithwaite, and Braithwaite, 2001; Moore and McDonald, 2000).

There remain difficulties in differentiating occupational violence and bullying, and in adjudication of less overt negative actions. The demarcation of meanings of occupational violence and bullying are discussed in the next sub-section. The application of a reasonable person test in adjudicating complaints about less overt forms of bullying/ violence is also considered.

Adjudicating meanings and reasonableness of occupational violence and bullying

While understandings of occupational violence and bullying have interpenetrated to the point that single definitions now address both concepts (such as the ILO *Code of Practice* discussed above), differences need to be acknowledged. As psychological violence, bullying is fraught with evidential difficulties, and attribution of cause and severity of effects differ between individuals and work cultures. This indeterminacy has been addressed by the introduction of a 'reasonable person' test into some definitions of bullying as a quasi-juridical criterion. For example, WorkSafe Victoria (2003: 6) defines workplace bullying as '...*behaviour that a reasonable person, having regard to all the circumstances, would expect to victimise, humiliate, undermine or threaten'*.

Incidents of overt violence are often one-off, clearly illegal, and produce predictable physical effects. However, traumatic effects of bullying accumulate from repeated exposure to variable low-grade negative actions, and fears thereof, over long periods of time (Leymann and Gustafsson, 1996). Thus, definitions of bullying more often refer to the *repetition* of incidents (Health and Safety Authority, 2002: 4). The nature and number of incidents over time necessary to qualify a recipient as having *experienced bullying* remains subject to debate (Leymann, 1997). Definitional terms and cut-offs can lead to different interpretations of the *experience* of bullying across the dimensions of 'exposure to', 'receipt of', 'experience of', and 'victimization'.

The discourses of bullying and mobbing have also opened a window into systemic factors that a *reasonable person* would consider a cause of work-related injury. As a consequence, work practices, deadlines, resourcing, training, and appraising that can be considered unreasonable have been named as both forms of bullying and risk factors therefore. For example, WorkSafe Victoria (2003: 6) treats '...*using a system of work as a means of victimising, humiliating, undermining or threatening'* as unreasonable behaviour.

Definitions by international peak bodies that address both bullying and violence are much needed as consistent points of reference for national agencies. However, it is important that such definitions address both common and disparate features of occupational violence and bullying. Arguably, the definitions by the ILO and the WHO presented above could be extended to acknowledge the desirability of a *reasonable person test* for bullying. The accumulation of post-traumatic stress effects from bullying through the progressive destruction of psychological integrity due to the *repetition* of (often minor) incidents (see: Chapter 5) could also be recognized. Furthermore, work systems and practices that a reasonable person would consider inappropriate or harmful could also be named as forms of bullying or systemic violence.

Forms and directions of bullying and violence

The matrix of forms and directions of occupational violence and bullying presented in Table 1.1 is very broad in depicting '*the thousand*

Table 1.1 **Forms and directions of workplace bullying and violence**

	Top-down internal	Lateral internal	Bottom-up internal	To clients	From clients	To external others	From external others
Physical Attacks	Assaults on staff by managers or supervisors Physical punishment, Torture	Staff-on-staff violence, retaliation, initiation rituals, or pranks	Assaults on supervisors or managers by employees Sabotage	Violence in eviction, repossession, debt collection Blocking access Unlawful detention	Attacks on staff, contractors or suppliers Destruction of corporate property Arson	Assaults on witnesses, community members, regulators, or other clients	Hold-ups Violent protest Vandalism Kidnapping Terrorism
Threats of violence	Threats of force in disputes with staff, contractors, or suppliers	Co-worker's threats of physical attack	Threats of violence to managers, supervisors, contractors or suppliers	Unreasonable warnings of forceful retaliation, eviction, or service termination	Threats of violent retaliation for poor service or product performance	Threats of physical harm to activists, or others in operating environments	Threats by robbers, extortionists, kidnappers, or terrorists
Verbal abuse	Yelling, swearing, name-calling and sledging, arising in conflicts over: expectations, entitlements, job conditions, rights, pay, work performance, management styles, work practices, interpersonal difficulties, and customer service or product quality						
Attacks on social status or position	Denial of participation in training or career opportunities; exclusion; sending to Coventry; blaming; scapegoating; demonising; denigration in terms of appearance, lifestyle, family, values, or ethnicity; vexatious allegations; exposure to malicious notes, rumours, or pornography			Denial of service because of characteristics, circumstances, or ethnicity	Attacks on the reputation of employees to gain trading advantage or service access	Appropriation of social or cultural, capital	Public humiliation by activists, competitors, or political rivals Identity theft

Table 1.1 **Forms and directions of workplace bullying and violence** – *continued*

	Top-down internal	Lateral internal	Bottom-up internal	To clients	From clients	To external others	From external others
Systemic violence	Unreasonable work practices, impossible deadlines, under-resourcing	Withholding information Interference with co-worker's files or equipment	Unreasonable work-to-rule Unfair refusal to operate new technology	Queuing or 'flicking' of complainants Impossible trading conditions Spam	Malicious disruption of corporate systems Fraud	Systems inflicting social and environmental damage Defamation actions to stifle protest	Computer hacking E-warfare E-terrorism
Epistemic	Conceptualizations of others that degrade their dignity and integrity and justify their maltreatment, for example, defamation, discrimination, race hate, or orientalism. Appropriation or unfair taxing of property. Intellectual suppression, e.g. misrepresenting, silencing, or excluding ideas or arguments			Fraudulent misrepresentation Untrue product claims Malicious rumours	Representations that degrade the dignity and integrity of employees	Propaganda for harmful lifestyles, products, or debt Green washing	Negative stereotyping Culture jamming Ideological conflict
Self-harm	Provocative behaviours that incite recipients to react aggressively. Risky lifestyles (e.g. extreme overworking, alcohol and illicit drug abuse, financial over commitment, or greed). Self-terrorisation; mental health problems; access to weapons; or suicide ideation. Complicity in unsafe work practices that give rise to risk of public backlash, loss of trading rights, bankruptcy, and civil or criminal charges.						

natural shocks the flesh is heir to' (Hamlet) in contemporary work-places. In addition to conventional understandings of bullying and violence, categories of *systemic violence, epistemic violence*, and *self-harm* are separated out. These categories may be conduits for bullying and violence that deserve specific attention in safeguarding an organ-ization. It may also be that prevention of occupational violence and bullying through one conduit leads to their expression through another.

Systemic bullying/violence

This categorization includes brutal work systems for which there are complex layers of responsibility. Early recognition that work systems or practices can be a *risk factor* for violence, and hence an OSH issue, has been evident in the discourse of occupational violence (Australian Institute of Criminology and The Division of Worksafe Health and Safety (Qld), 1993) for over a decade. Precursors for the categorization of unreasonable work systems and practices such as mobbing (and bullying) can also be found in the Swedish National Board of Occu-pational Safety and Health's Code *Victimisation at Work* (1993). Since then, work systems or practices that place employees, clients and members of the public at risk of psychological injury have been termed *organising-violence* (McCarthy *et al*, 1995), *systems bullying* (Kennedy, 2001) or *systemic violence* (Bowie, 2002).

Where otherwise 'normal' individuals occupy roles that relay bully-ing and violence, in the 'way things are done around here', the safe-guarding challenge is formidable. Trials of the perpetrators of the Holocaust and of the genocide in Yugoslavia in the 1990s testify to vi-olence inculcated in system roles. These trials showed that remedies for gross violence perpetrated in the name of organizational roles and mis-sions are possible where there is an international system of justice that has wide subscription from nation states (see: Chapters 5 and 9–10).

Epistemic (or conceptual) violence

This category recognizes the potential for the manipulation of mean-ings to be implicated in the perpetration of bullying/violence. Conceptual violence can be perpetrated through fraudulent misrep-resentation, defamation, discrimination, race hate, and orientalism. For example, restructuring, downsizing and the relocation of commu-nities can be legitimated by the negative stereotyping of the 'un-productive' in the interests of 'progress'. Such denigration also has its counterpart in the urban renewal that gentrifies and gates communities in the name of rectifying obsolesce and enhanced security (McCarthy, 1998). The use of epistemic violence in a ritual sacrificial manner in rationales for the cleansing of work and urban environments for the 'greater good' is discussed in Chapter 5.

Defamation is a core tactic used by perpetrators of epistemic violence, and remedies can be sought through criminal and civil laws. However, the prohibitive cost of legal action to remedy defamation makes this pathway inaccessible for most. Remedies for damages occasioned by such misrepresentations can be sought through organizational complaints, conflict resolution or grievance procedures, as well as tribunals and ombudsmen. However, complaints are often difficult to evidence, or to distinguish from 'reasonable management practice'. The allegation that a person is a perpetrator of aggression also raises risks of defamation, necessitating protection for complainants (see: University of Queensland, 2003).

A tendency for managers who face allegations of bullying/violence to react by initiating performance appraisals and disciplinary proceedings against complainants has been evident in some recent case material (McCarthy and Mayhew, 2003). Here, misrepresentation can arise in counter-allegations that the complainant's performance was in some way 'unsatisfactory', or constituted 'misconduct' and that reasonable management action had been exercised. Such counter allegations can be tactical to weaken the grounds of complaint, and can also serve as an ingratiating signal to power-holders threatened by allegations of impropriety.

Self-harm

The category of self-harm has been included in Table 1.1 to accommodate the potential for individuals in their organizational roles to variously internalize, relay, or be complicit in the transmission of bullying/violence in ways that may be harmful to them. As such, the concept opens a space for discussion of complicity and self-responsibility. Recipients of bullying/violence can internalize pain in unhealthy victimhood, for example in apathy, hopelessness, depression, obsessive pursuit of justice, revenge, and suicidal ideation. Recognition of the potential for self-harm brings us face-to-face with our complicities, as victims, witnesses, relays, or perpetrators and gives rise to questions of responsibility for others and ourselves. As such, the notion of self-harm prompts thought about what might be entailed in moving from unhealthy victim identities, or relay positions, to becoming agents for change (McCarthy, 2001, 2003a).

Risk factors

The identification of risk factors is a key endeavour in the chapters that follow. Findings from new research studies completed in the retail, transport (including taxi, courier and long-haul truck drivers), health, seafaring, tertiary education and juvenile justice industry sectors provides a basis for identifying and profiling risk factors for occupational violence and bullying (see: Chapters 4, 6, 8). In the small business retailing sector, the fear of hold-ups was belied by low exposure to actual incidents, and, although shop-lifting and credit card fraud were commonly experienced, the motivation for their prevention was found to be low amongst owner managers. However, intimidation and harassment from youths loitering around retail premises was identified as a significant concern, with a potential to frighten away customers.

Findings from the studies completed in the health, seafaring, transport and juvenile justice sectors align occupational violence and bullying on continuums that graduate the severity of perpetrator *actions* from the severity of *impacts* on the recipient (Mayhew and Chappell, 2003). The *severity of actions* was differentiated on a continuum spanning homicides, physical violence, threats of violence, verbal abuse, bullying (including mobbing) and other actions that do not fit neatly into the categories (e.g. spitting at staff). The study findings suggest that employees in these industries routinely experience significant levels of occupational violence and bullying.

The interaction of risk factors in the escalation of risk of frequency and severity of events and their impacts are also key discussions throughout this book. Common formations of risk are found in a field of tensions between managerial, systemic, community, socio-economic, governmental, cultural, and global forces (see: Chapters 2–5 and 9). The spillage of bullying and violence from within workplaces into local communities is found to fuel resentment and seed further aggression that blows back on workplaces. The notion of 'psycho-terror' (Leymann, 1990) provides a bridge between occupational violence, fear of aggression and global terrorism that prompts the identification of mutually furthering moments (see: Chapter 5).

Modelling of costs

Estimation of the costs of occupational violence and bullying has played a key part in advocacy to enhance prevention by managerial and governmental bodies. The modelling of costs in Chapter 3 broadens the domain of concern beyond direct productivity, staff turnover and health effects. Costs that are less tangible include damage to reputation, an enabling environment for corruption, and degrading of individual rights in social democracies. Costs of bullying and violence are shown to accrue at individual, organizational and wider social, economic, civil, political and cultural levels. The need to account for the costs of highly unlikely events (e.g. massacres or terrorism) which can severely damage the enterprise (see: Chapters 5 and 9) is also acknowledged. The investment in safeguarding an organization from the costs of bullying/violence is considered justified by the magnitude of the benefits to be realized.

Pathways to safeguarding

Scientific research studies have provided the basis for the recommendation of industry-specific safeguarding strategies (see: Chapters 4 and 6–8) that can be fine-tuned for application in most industry sectors. For example, the *Crime Prevention and Environmental Design* (CPTED) strategy outlined is proffered as a necessary component of the strategic response in all workplaces (see: Chapters 4 and 6). The *Crime Prevention in Small Business: Self Audit Checklist* (see: Appendix 1) can also be adapted for use in medium size and larger organizations. Many of the strategies recommended in these instruments are also likely to mediate the escalation of bullying into overt violence.

The administrative interventions discussed in the chapters are based on scientific research evidence drawn from industry. Our recommendations include the implementation of zero-tolerance policies, tailored prevention strategies, codes of conduct, emergency response planning, as well as the mainstreaming of safe work practices, training, reporting, performance appraisal and evaluation. Six critical stages through which risk and severity of workplace bullying and violence typically escalate are modelled in Chapter 9. Checklists for the *Audit of Safeguarding Capacities* by employee, managerial, industry, and governmental interests are presented. These audit checklists have been compiled to motivate cross-sectoral responses and improve safeguarding.

The road map to safeguarding presented in Chapter 10 outlines a basis for an international response. The approach is based on steering diverse disciplinary, professional, industry and governmental interests towards an international protocol addressing workplace violence and bullying that has widespread acceptance by national governments. The case for realization of mutual benefits for governments and organizations that subscribe to the protocol is made in the interests of sustainable safeguarding.

Conclusion

The multiplicity of meanings, forms, directions, and interactions of occupational violence/bullying pose a formidable challenge for safeguarding. New formations of risk have emerged from the interplay of global and local forces in which risk and severity incubates and escalates. Significant levels of exposure to occupational violence/bullying have been found in new industry sector research. Modelling of the costs of bullying and violence has identified that considerable financial, productivity, health and well being, reputational and market benefits can be derived from safeguarding. An integrated business, community and government approach over the longer term is considered necessary to meet the complexity of the safeguarding challenge. The approach recommended entails the generation of shared interests and goals across employee, managerial and governmental interests so as to motivate weaving of current disciplinary and sectoral approaches into a web of cross-sectoral safeguards. The orientation of safeguarding programs to international best practice is important to improve the viability of preventative strategies within national jurisdictions. A global anti-workplace bullying/violence initiative is necessary to achieve the critical mass of support for the development of best-practice standards that have wide acceptance by business, governments, and employees.

2

The Overlaps Between Occupational Violence/Bullying and Systemic Pressures on Organizations from Global Market Environments

Claire Mayhew

The inappropriate behaviours adopted by perpetrators of occupational violence occur along a *continuum of physical severity,* and range from verbal abuse, ridicule, bullying, sexual attacks, and threats to assaults. That is, the violent activities range from insensitive verbal interactions or unreasonable demands to criminal acts (Mayhew, 2002a).

Conceptual typologies

In order to better understand the manifestations of the different forms of occupational violence, there are various conceptual typologies that can be adopted to enhance understandings.

First, it is common practice among those who work with perpetrators of violence in the general community to differentiate events by the *severity* of the physical aggression. Hence, whether or not the violent event is encompassed within criminal codes is of central importance in this conceptual approach.

Second, and commonly following on from the above point – characteristics commonly identified among *perpetrators of violence* can be the core differentiating variable during analysis. For example, young males are almost universally over-represented among perpetrators of violence in the community (see: Chapter 4; BCS&R, 2003).

Third, the recent rise of terrorism directed to worksites where there are multiple potential victims is a recent phenomenon that has sparked national initiatives across industrialized countries (see: Chapter 5). Many of the preventive approaches to date have been based on identification of the characteristics of high-risk potential perpetrators;

for example, particular nationalities may have been singled out for more in-depth assessment at many international airports.

Fourth, occupational violence can be differentiated on the basis of the *severity of impact* on recipients. This type of conceptual approach is in its infancy. However, as research evidence accumulates it may be possible to more accurately differentiate *emotional injury/stress outcomes* from the physical severity of the event using this conceptual approach. The early evidence that can be included within this conceptual approach indicates that events at both ends of the *continuum of violence* can result in significant emotional injury. That is, the *impact* of occupational violence may be unrelated to the physical severity of events and more closely correlated with the presence/absence of malice by the perpetrator (see: Mayhew *et al*, in press 2004).

Fifth, occupational violence may be conceptualized on the basis of *preventability*. In such an approach, it must be possible to formally evaluate the relative effectiveness of diverse interventions (Mayhew, 2003b). Arguably, this type of conceptualizations offers the greatest preventive capacity in the longer-term, although accurate data about incidence and severity, high-risk scenarios, and perpetrator characteristics are essential pre-conditions.

Sixth, the most commonly adopted basis for assessment of occupational violence is an approach based on occupational violence/bullying patterns identified in databases. The obvious weakness with this approach is that formally reported data may not accurately represent the full range of violent events that occur. For example, the international research evidence indicates that, at best, only around one in five violent incidents are formally reported (Mayhew and Chappell, 2003; Chappell and Di Martino, 2000). Non-reported incidents are commonly referred to as the 'dark' figure of occupational violence (Mayhew, 2002a).

Seventh, a typology of aggression at work was developed by the Californian Occupational Safety and Health (OSH) authority which has been widely accepted by OSH authorities across the world, the criminal justice system and among researchers. The CAL/OSHA (1998) typology separates occupational violence into three basic categories, based on the relationship of perpetrators/offenders to the organization where the inappropriate behavior occurs.

- 'External' violence is perpetrated on workers by persons from outside the organization, such as during armed hold-ups at shops, service stations and banks;

- 'Client-initiated' violence is inflicted on workers by their customers or clients, such as patient aggression towards nurses; and
- 'Internal' violence is inflicted by one worker in an organization on another, such as between a supervisor and employees, or workers and apprentices, and may include initiation rites, assaults, or bullying (CAL/OSHA, 1998).

The perpetrators of the different forms of occupational violence in the CAL/OSHA typology have distinct characteristics, and the most appropriate prevention strategies differ markedly between the different forms of aggression at work.

More recently, it has been proposed that there is a more 'systemic' form of occupational violence which arises from widespread economic and social pressures. 'Systemic' violence may arise from excessive economic stress or productivity demands, and results in work intensification, job insecurity, heightened anxieties among the workforce, excessive hours of labour, and may contribute to a workplace culture where threatening behaviours are tolerated (Bowie, 2002; McCarthy *et al*, 1995).

These different ways of conceptualizing occupational violence need to be interpreted as 'ideal types' that do not necessarily occur separately. For example, all the different CAL/OSHA forms of occupational violence can occur on the one site at different times, or arise under a range of pre-cursor scenarios. That is, an intoxicated patient may abuse a nurse, and then later threaten an X-ray technician with assault if procedures are delayed, or may even attempt to 'hold-up' the pharmacy for drugs. Similarly, there is a close overlap between 'internal' occupational violence/bullying and the 'systemic' form. 'External' violence (in the CAL/OSHA typology) can also be perpetrated by bystanders such as when ambulance officers are assisting the injured at pub brawls. Further, stressed employees can direct aggression to clients as well as colleagues. Or, the recipients of 'internal' violence/bullying/initiation rites may adopt the behavioural pattern directed to themselves and become aggressive to third parties. A more detailed overview of the different forms of occupational violence – as conceptualized by CAL/OSHA – appears below, together with the known risk factors.

'External' occupational violence

'External' violence is perpetrated by offenders from outside the organization who are seeking cash or highly-valued goods, for example, during an armed hold-up at a 'corner shop', a service station or a bank.

Unsurprisingly, the jobs at highest risk of 'external' violence are those where significant amounts of cash are on-site. The high-risk premises and occupations for 'external' occupational violence are similar across industrialized countries. Workers in banks, post offices, gambling outlets, armoured vehicles transporting cash, taxi drivers, 'convenience' stores, liquor stores and garages are at increased risk of 'external' occupational violence almost everywhere (Chappell and Di Martino, 2000; Trades Union Congress (TUC), 1999). There are four core business risk factors for 'external' occupational violence: cash/valuables on-site, few workers employed at any one time, evening/night trading, and face-to-face contact between workers and their clients/customers (Mayhew, 2002a; Heskett, 1996). Repeat victimization is common in some geographical areas ('hot spots'), and little occurs in others (Fisher and Looye, 2000). Bellamy (1996) explained this trend in terms of 'attractive' versus 'unattractive' targets. 'Attractive' targets were situated in high-crime areas, had few preventive interventions, provided for limited observation from passers-by, allowed quick access to highways for get-away, and had a number of potential exits from the site (OSHA, 1998).

   The targets of 'external' violence also shift over time as offenders seek out premises with limited security that have lots of cash or highly desired goods (known as 'hot products') on-site. A 'hot product' is one that retains a high market value and is readily re-sold, for example, laptop computers, name-brand sports clothing, alcohol or cigarettes (see: Mayhew, 2001a for a detailed discussion of cargo theft). Further, as larger organizations improve their prevention efforts, perpetrators may seek out 'easier' targets, such as 'corner' stores or video outlets. Taxi drivers are at particularly high risk across most if not all industrialized countries as they are relatively 'easy' targets vis-à-vis financial institutions which have progressively improved their security measures (Mayhew, 2000a). Similarly, an exponential rise in 'external' violence has also been noted for chemist shops as offenders seek pharmaceutical as well as 'easy' cash (Taylor and Mayhew, 2002; Peronne, 2000). Small businesses whose owners/managers cannot afford expensive preventive interventions may also be at increased risk of 'external' occupational violence (see: Chapter 4). For example, an Australian study of small retail shop owner/managers reported that 4% experienced a hold-up over a 12-month period (Mayhew, 2002b). These hold-ups occurred in addition to other forms of crime victimization, including: shoplifting (76%), intimidation/harassment (58%), break and enter when closed (42%), credit-card or cheque fraud (42%), employee theft (36%), other fraud (24%), and other problems (18%) (ibid). That is, the total burden from crimes against small business owner/managers can be significant. Nevertheless, the rewards to the (often desperate) perpetrators can be quite small. The motivation factors are quite different for 'client-initiated' violence.

'Client-initiated' occupational violence

'Client-initiated' violence is perpetrated by the customers or clients of a service-based organization. The jobs at particular risk of 'client-initiated' violence in the US, Britain, and Australia are: police, security officers, prison guards, fire service, teachers in primary and secondary schools, health care and social security workers (Mayhew, 2002a; Fisher and Gunnison, 2001; Chappell and Di Martino, 2000). At particular risk are workers in jobs where they: '... provide care and services to people who are distressed, fearful, ill or incarcerated' (Warshaw and Messite, 1996: 999).

Similar risk factors and high-risk scenarios have been identified in different studies conducted in various countries; a brief summary of which is now provided. Clients who are intoxicated with alcohol or illicit substances have been reported to pose an increased risk (Fisher *et al*, 1998). Correctional and juvenile detention workers are at increased risk from young males with a past history of violence (Flannery, 1996). Younger males who suffer psychosis or a neurological abnormality also pose a greater risk to staff caring for them (Turnbull and Paterson, 1999). Research conducted in the Australian health care sector also reported that the perpetrators of 'client-initiated' violence were disproportionately male, younger, affected by substances or suffering from dementia (Mayhew and Chappell, 2003; Brookes and Dunn, 1997). (Client-initiated violence in the health industry will be explored in depth in Chapter 6, and the risks involved with educating and caring for adolescents are analysed in Chapter 8).

Across industrialized countries, the incidence and the severity of 'client-initiated' violence appears to be increasing. The causes for this increase in aggression are complex and include (among other things), a general move to 'user pays' systems for a range of services, cutbacks in publicly funded supports (such as free health care, public housing etc), and a polarization of the population in many countries towards a smaller proportion of very wealthy individuals and a larger group of welfare dependent or 'working poor' families/groups. Economic stress can be exacerbated through unemployment, work casualization, restricted access to unemployment benefits, or ill-health. When economic stress occurs in a context where many others live in relative affluence, aggression may be directed towards staff members who restrict access to support services, where there are long waits to see health care workers, or access to public housing is tightly rationed. When cuts in public spending and services occur, the clients may respond violently believing they are being treated unjustly or unfairly (Mullen, 1997). In contrast, the third type of occupational violence identified in the CAL/OSHA (1998) typology arises in quite different scenarios and from perpetrators with distinct characteristics.

'Internal' occupational violence/bullying

'Internal' violence is inflicted by one worker in an organization on another, such as bullying between a supervisor and a subordinate, a worker inflicting initiation rituals on an apprentice, or through 'bastardization' rites in the armed services. That is, supervisors or co-workers are the perpetrators of these forms of occupational violence. An international study estimated that at least 10% of European workers were being subjected to 'internal' occupational violence/bullying of one sort or another from colleagues each year (Hoel *et al*, 2001). The majority of incidents reported in studies are 'top down' in direction, although lateral and 'bottom up' violence/bullying behaviours are also common (see, for example, Mayhew and Chappell, 2003; McCarthy *et al*, 1995). Thus, with 'internal' occupational violence/bullying perpetrators and recipients are known to each other, with many having daily contact. Some of the contexts and characteristics associated with particular forms of occupational violence/bullying are detailed below.

Bullying most commonly emerge in organizations where dominant/ subordinate hierarchical relationships exist, such as with quasi-military supervision arrangements (McCarthy *et al*, 1996). Many of the tactics used by perpetrators are initially subtle and covert, but intensify and escalate to more overtly demeaning and threatening behaviours over time. Similarly, sometimes poor management techniques progress and evolve slowly over time (McCarthy, 2002; Keashley, 2001; Chappell and Di Martino, 2000; Workers' Health Centre, 1999). The *bullying* incidents are usually repeated, and escalate in intensity over time. When the victim/recipient leaves the workplace, another victim is usually selected (Mayhew, 2000c). Further, a 'culture of denial' can permeate the organization whereby victims – rather than perpetrators – are blamed. The perpetrators of bullying can be motivated by a wide range of personal inadequacies, including envy of more popular or skilled employees, positive consequences from adopting bullying as a strategy while at school, or these offenders may be unable to control their aggressive tendencies (Gaymer, 1999; Warshaw and Messite, 1996). The evidence is now overwhelming in that bullying may result in serious emotional consequences for the recipients/ victims (Mayhew *et al*, in press; Mayhew and Chappell, 2003; McCarthy, 2002).

Degrading *initiation rites* inflicted on young new workers have been reported in a range of industries (McCarthy, 2002; Chappell and Di Martino, 2000). In some organizations, long standing employees may have accepted initiation rites as 'normal', such as in some units of the armed services. Similarly, initiation rituals have been common- place for beginning apprentices in a number of places. On occasion, these initiation rites have escalated into full-blown violence, and, as a result, there have been a series of prosecutions of both supervisors and co-workers who harmed the victim/recipient. Nevertheless, in the majority of instances, overt violence is uncommon.

Threats and physical assaults are also less common in most industrialized countries than are bullying or initiation rites. Most overt 'internal' occu- pational violence is preceded by warning signs such as belligerent or intimidating behaviors (Chappell and Di Martino, 2000). Nevertheless, the US has experienced a series of mass shootings from disgruntled or dis- missed employees who had previously engaged in threatening behaviour. The risk factors include: being sacked, imminent retrenchment, long-term employees who perceive that they have been treated unfairly, older males with families to support, and intoxicated young males (Standing and Nicolini, 1997; Capozzoli and McVey, 1996; Myers, 1996). Yet relying on profiles is risky as perpetrators have been identified who have a wide range of grudges and motivations and diverse characteristics.

Organizational features: It is important to recognize organizational contexts and cultures where these 'internal' forms of occupational vi- olence/bullying more commonly arise. A quasi-military hierarchy, rigid management styles, job insecurity, overwork, management toleration of initiation rites, a culture that ignores bullying, marked divisions between supervisors and employees, 'normalization' of practical jokes, and a highly competitive business environment all increase the risks (McCarthy *et al*, 2001; Mayhew, 2000c; Mullen, 1997). Such organiza- tional characteristics can also change over time as pressures in the broader economic environment alter.

'Systemic' occupational violence

It has been argued that 'systemic' violence arises out of broader eco- nomic forces in society, particularly when inordinate production demands are placed on organizations and workers. The consequences may include any of the following: work intensification, cost-cutting, over-work, long shifts, tense inter-personal relationships, toleration of aggressive behaviours towards and between workers, and the development of a culture where a range of forms of occupational violence/bullying are tolerated.

Essentially 'systemic' violence arises and is sustained where global market capitalism and post-modern culture combine and are translated internally into a violence-conducive organizational culture. McCarthy *et al* (1995) highlighted the 'inappropriate coercion' that commonly surfaced in organizations undergoing chronic restructuring. These authors identified a historical lineage of aggression which they termed 'organizing violence' that was triggered by excessive pressures and global market crises, resulting in stress, inappropriate poor people skills, bullying and enhanced fears among the workforce.

Violence-conducive cultures implicitly value a propensity to aggression through rewards systems, and devalue any organizational protocols that mitigate such activities. Foucault (1975) described this type of dynamic as the 'contours and flows of power' by which rampant economic and cultural forces flowed into workplaces. In cultures where the pursuit of accumulation through global market capitalism is valued, where violence in sport and entertainment is normalized, and where there is widespread acceptance of overwork, an organizational culture of *'systemic'* violence can readily arise (Hatcher and McCarthy, 2002). The underlying pressures are complex and involve economy, government, global capitalism and post-modern culture.

Yet, arguably, external economic pressure supported by a societal value system that lauds excessive competitiveness are not sufficient on their own to lead to occupational violence. These economic exigencies and values must be incorporated within organizational systems to unleash the 'systemic' violence/bullying propensities of individuals. The remainder of this chapter is focused on elaborating how these external 'systemic' pressures are adopted and are translated by those working within an organization into 'systemic' occupational violence.

The translation of systemic pressures from the external environment into 'systemic' occupational violence

It is argued that external pressures, organizational culture, and individual behaviours interact in a complex and multi-dimensional manner to create conditions where 'systemic' violent propensities can flourish. Liefooghe and Davey (2001) have argued that both *pathological organizations* and *pathologised individuals* are essential for the maintenance of a harmful workplace culture that is dominated by a negative external environment. Similarly Zapf (2001) has argued: 'Leadership problems and organisational problems cannot "harass" an employee. Such behaviour is only possible for human beings'.

In this chapter, it is argued that systemic violence is manifest only when two essential pre-conditions exist:

(a) A hostile external environment (usually associated with, and propelled by, significant economic stress); and
(b) An internal *conduit* person who accepts and then transmits these pressures into excessive and unreasonable demands on other workers.

In essence, the perpetrators become *conduits* passing external pressure onto subordinates. Some perpetrators will be *intentional* conduits of *systemic* violence (or even enjoy enacting sadistic behaviours), while others may be coerced progressively over time through threats to their own economic survival. Yet, arguably, few will resist the pressure to adopt inappropriate behaviours in an organization where systemic pressure is mounting. For example, senior staff members (or co-workers) may adopt a range of 'culturally acceptable' behaviours when faced with rampant competitiveness from the external marketplace; only some of their resulting inappropriate behaviours may lie within the ambit of occupational violence. Sometimes the demands placed on workers will be unreasonable only in the short-term; while in other places behaviours will clearly constitute sustained bullying/overt violence. For many *conduit perpetrators* of systemic violence, patterns of inappropriate behaviour are thrust upon them with many becoming more or less complicit because they need to maintain their employment and income. As has previously been identified with bullying, the behaviour adopted as a consequence of systemic pressures are also usually repeated and escalate in intensity over time, particularly when the external pressures on perpetrators are sustained (Mayhew, 2000c; McCarthy *et al*, 1995).

It is argued that there are ten basic strategies adopted by these perpetrator conduits of systemic external pressures, all of which can be located on a 'continuum of malevolence'. At one end of the continuum are the individuals who 'look elsewhere' when inappropriate behaviours occur, to those at the other extremes who enjoy sadistic behaviour.

The strategies adopted by conduits described in this chapter are 'ideal-type' assemblages of unreasonable behaviours that have been observed to manifest when wider economic and cultural forces are translated and become embedded within organizations. This proposed typology is based on real people working in the health care, university, and civil service industry sectors in Australia and Britain, although names have been removed and organizations de-identified. These roles are not static, and elements of each of the behavioural assemblages may emerge in particular persons at different points in time, under specific circumstances and as people interact with one another. Each person may transmute from one form of behaviour to another as they move between organizational roles, and as opportunities to enact latent pathological leanings arise.

The ten basic behavioural strategies that can be adopted along the 'continuum of malevolence' when systemic environmental pressures rise rapidly include: (a) the *missionary conduit* (b) the *apathetic* (c) the *busy boss* (d) the *egomaniac* (e) the *Lady MacBeth* (f) the *Rottweiler* (g) the *strategic* bully (h) the *Mr Mafioso* (i) the *career bully* and (j) the *industrial psychopath*. The typical behavioural manifestations and tactics adopted by each of these 'ideal types' is detailed below.

The 'missionary' conduit

This type of supervisory conduit may be altruistic, focused on 'good works' in the community, hard working and may genuinely not have the time to address inequities that arise in any organization from time to time. Precisely because this 'missionary' conduit is so committed to his/her 'good works', other issues become secondary considerations; hence s/he abdicates responsibility for the human sacrifice made by others. For example, inappropriately heavy workloads may be displaced onto subordinates. Sometimes a rationale will be used (and believed) to explain away inequities; for example, 'the recipient is very competent and is being tested for future advancement'. In such scenarios, the resulting overwork is unlikely to constitute occupational violence under any present definitions. Importantly, *malice* is usually totally absent among these 'missionary' supervisors who transmit excessive demands onto colleagues and/or subordinates. The reasons why recipients accept the arrangements and allow them to persist over time may include powerlessness in the labour market, a sharing of the goals of the 'missionary', or a degree of masochism. Nevertheless, these exploitative employment relationships are likely to be of particular interest to those involved with labour law, equal opportunity and OSH.

The 'apathetic' conduit

This type of supervisory or co-worker conduit is generally struggling to cope with the demands placed on him/her by the external market-place, has trouble adapting to changes in production methods or products, is frightened of innovation, is imbued with following work practices in time-honoured ways, and is clinging onto a job in order to maintain an identity, lifestyle or consumption pattern. Such 'apathetic' conduits rarely recognize that they are displacing production demands or requirements for innovation onto others, or that historically 'normal' behaviours and interactions may constitute inappropriate behaviour under new regimes. Typically, these 'apathetic' conduits view themselves as a member of the 'silent majority' as they resist change on a range of fronts. For example, in workplaces where a co-worker is being sexually harassed, bullied, or subjected to excessive production demands, an 'apathetic' conduit will ignore the recipient, and become complicit by default. Nevertheless, he or she generally recognizes at some level that they are an historical anachronism in a global market economy. That is, while an 'apathetic' conduit is rarely deliberately malicious, he or she will and try to protect themselves at any price while maintaining a pleasant inter-personal manner.

The 'busy boss' conduit

The 'busy boss' type of supervisory conduit is also usually very hard working and committed. However, in the face of unrelenting work pressures, commitment to 'higher order' goals supersede the addressing of inequities in workload, inappropriate resource allocations, or care of other workers. Unlike the 'missionary' conduit, business survival is the aim of this supervisor, rather than altruistic outcomes. Again, pressures displaced onto subordinates are unlikely to be considered 'violent' under current definitions. However, some of these supervisors may grow into bad interpersonal habits over time e.g. inappropriate coercive behaviour at times when production demands are high. Inappropriate interpersonal communications can also be the outcome of excessive stress, or develop from a 'bad management' belief that pressure 'gets the best' out of workers (Mayhew, 2002a).

The conscious tactical decisions made by 'busy boss' conduits may be overlooked by recipients, as may the *structural* forces that underlie these scenarios. For example, many recipients of inappropriate behavior may excuse the perpetrator by focusing on the conduit's personal inadequacies in coping with the heightened pressures. For example,

McCarthy *et al* (1995) found that 70% of the recipients of bullying attributed these inappropriate behaviors to a manager's 'poor communication skills'. Nevertheless, as Mumby and Stohl (1996) have argued, traditional styles of management *require* the exercise of power and this in itself can inadvertently promote tyranny. Ashforth and Humphrey (1995) have also elucidated some of the strategies that may be adopted in the exercise of 'petty tyranny'. Most of these perpetrator conduits are, arguably, unaware of the severe impact on recipients as they try to function at the limit of their skills – which may lead to frequent lower-level aggression such as widespread insensitive negative verbal comments.

In all of these 'busy boss' scenarios, the CEO is likely to have heard that inappropriate behaviour is occurring, but may have chosen to do nothing. The CEO may believe that the victim is to blame in some way, delude him/herself that the perpetrator is a very productive supervisor/employee and just has some 'minor' faults. Alternatively, the CEO may engage in 'wishful thinking' that this problem will just 'go away'. Or, the CEO may him/herself be the perpetrator.

The 'egomaniac' conduit

This type of conduit is focused on his/her *personal* career or pay advancement and will not be deflected by excessive demands from the external environment, the well being of other members of the work group, equity, or care of the recipients of inappropriate behaviour. Only those external threats and problems that can be manipulated or adapted for personal progression will be taken up. For example, an 'egomaniac' supervisory conduit may refuse holiday leave applications from subordinates at high-demand times to ensure others do unwanted tasks, may 'forget' to forward on advantageous information, may reserve interesting tasks for him/herself, or may gently denigrate co-workers' abilities to cope. If the recipient remonstrates, some sort of 'personality conflict' or inadequacy in the subordinate may be blamed. Another variant of behaviour adopted by one 'egomaniac' conduit who was an on-site union delegate in a rapidly downsizing public service department was to immediately inform management of confidential union decisions. (However, since his duplicity was known to other delegates, misinformation was deliberately discussed at all meetings where the 'traitor' attended.) At the same time, most 'egomaniac' conduits are interpersonally skilled, have a high public profile, and are charming and popular at a superficial level. This 'egomaniac' conduit is qualitatively different to the 'busy boss' supervisor in that *personal* progression rather than business survival is paramount. Overt conflict very rarely appears. Hence malice is usually absent – although deliberate shifting of unattractive tasks, gross overwork, or redundancy risk to others is the norm.

The 'Lady MacBeth' conduit

The 'Lady MacBeth' conduit usually operates surreptitiously and via a 'front man' who does the 'dirty work'. A Lady MacBeth conduit is focused on removing potential competitors *before others recognize them as such*. When a threat appears on the horizon, unspoken assumptions among co-workers may be *quietly* played upon. For example, in a male-dominated workforce in a competitive economic environment, there may be implications that female job applicants 'haven't got the balls' for a difficult task. Other character assassinations may include comments about the recipient's childcare needs, sexuality or political affiliations. The result is that potential competitors are removed from a short-list before it is even written. 'Lady MacBeth' conduits are normally *deliberately but covertly* malicious. Hence it is usually difficult to track the consequences of this 'systemic' occupational violence directly to the perpetrator.

The 'Rottweiler' conduit

The 'Rottweiler' conduit adopts a very aggressive inter-personal manner in the face of unrelenting and systemic external environmental pressures. Typically, such people have been hired to perform roles where assertive verbal interactions are prized e.g. in industrial relations negotiations or in court. Such verbally aggressive interactions may become 'normalized' in their everyday manner throughout both busy and quiet times, and with both adversaries and subordinates. However, subordinates (particularly those new to the job) may be significantly emotionally injured by this 'systemic' behaviour. These 'Rottweiler' conduits are rarely deliberately malicious with co-workers and subordinates, although extensive harm may occur. Notably, many 'Rottweilers' are oblivious to their own excessive aggression and its impact on recipients.

The 'strategic bully' conduit

The aim of this supervisory (or co-worker) conduit is to protect themselves at all costs, including career advancement, reduction of workload, and personal progression. The 'strategic bully' conduit is distinct from the 'busy boss', 'egomaniac', and 'Lady MacBeth' conduits in that the aggressive behaviours are vibrantly overt and are deliberate. As external environmental pressures increase and overall workloads go up, the 'strategic' conduit will shift work to others, and yet at the time when job tasks are nearly complete, personally claim ownership of productivity or intellectual property – and denigrate or exclude those who remonstrate.

The 'strategic bully' conduit flourishes in a more autonomous environment (such as a university) where s/he can avoid collaborative work and less attractive tasks through being grossly objectionable. There is a range of strategies that can be adopted by 'strategic' conduits including overtly aggressive behaviour, demeaning of others, avoidance of any additional tasks, deliberate assertion of offensive views, and/or deliberate use of *negative* interpersonal 'skills' to ensure that no one wants to work with him/her. In initial stages of inappropriate behaviour, activities may occur one-on-one and out of sight of others. Belittling may be masked as 'helping' subordinates or co-workers to more competently do their job, such as through invasion of personal space, stealing of intellectual property, or a cycle of praise and blame adopted with subordinates with the aim of unsettling the recipient so s/he never knows where s/he stands. Over time, one-on-one behaviours become more overtly negative in public and may be in the guise of a joke that denigrates the recipient (or the group/race/gender the victim belongs to) in some way. This *repeated* behaviour is likely to unsettle the recipient so that s/he is not performing to previous standards. The decline in the productivity of the recipient is likely to be seized on by the perpetrator. That is, cause and effect can become confused. There is now a rationale for the perpetrator to be awarded additional resources, to be promoted, to publicly exaggerate the shortcomings of the recipient, and to concomitantly enhance his/her own attributes with impunity.

Co-workers are likely to be aware that something rather 'nasty' is occurring from the beginning, but many witnesses may be terrified that they will be the next recipients and so they 'keep quiet' and turn away from the victim. Research studies report that this is often the most hurtful of all behaviours for victims – who are now isolated (Mayhew and Chappell, 2003; McCarthy, 2002, Einarsen, 2000a and b). Malicious rumours may even be spread with impunity. Notably, 'strategic' conduits regularly use malice as a weapon as they translate systemic pressures into 'systemic' occupational violence.

The '*Mr Mafioso*'

This type of conduit holds *implied* power, in addition to that associated with their formal work role. A '*Mr (or Ms) Mafioso*' conduit is likely to imply that he or she has links with intelligence agencies, crime figures, or people who have unusual physical prowess. Sometimes, the '*Mr Mafioso*' makes no reference to family connections in the underworld, but allows this information to 'leak out'. Thus, this form of conduit quietly plays on the underlying fears or vulnerabilities of coworkers or subordinates. Typically a '*Mr Mafioso*' conduit is in a position of power, but feels insecure in this role. For example, he or she may have been appointed to their position as a compromise candidate, been assisted by equal opportunity provisions, or hired because no one else wanted to do a difficult job. That is, a '*Mr Mafioso*' conduit will adopt threatening behaviour – or even illegal means – whenever or wherever he or she feels it necessary to dampen opposition or to prop up their own power base. Thus, ethical standards of behaviour are irrelevant, except on paper.

The 'career bully' conduit

Many of these conduits of *systemic* violence adopted bullying as a strategy at school and found it 'worked' for them. As an adult, a practiced bully may be hired to do unpleasant tasks, for example, to encourage a number of staff to resign or accept redundancies, for example, in a downsizing public service department. McCarthy *et al* (2001) have previously identified that the emergence of workplace bullying is facilitated at a time when there is both increased insecurity and additional demands on employees, such as during restructuring of an organization. One important caveat vis-à-vis other *systemic* conduits is that career bullies generally 'enjoy' their role and gain the satisfactions and rewards from the resulting promotions etc.

Typical strategies adopted by 'career bullies' include: making workers appear incompetent through assignment of meaningless tasks, isolation of recipients, blocking promotion, taking credit for other's work, 'game' playing, threats to employment security, work overload, and practical jokes (Mayhew and Chappell, 2003). In the early stages, it is often hard to prove that inappropriate behaviour is occurring, as many of the tactics are *covert* and aimed at undermining the recipient. For example, errors of omission may occur as the supervisor 'forgets' to provide essential information or resources to do a required task. Or, abusive e-mails (known as 'flaming') may be repeatedly sent. The tactics then evolve and may include public demeaning, throwing furniture, or stealing documents from files held by others (Mayhew, 2000c). A common tactic adopted by career bullies is to very negatively evaluate the work quality and productivity of recipient/victims during formal annual performance appraisal procedures. Because these performance appraisals are generally required under enterprise agreements, and since supervisors have a *legitimate* role in evaluation, the recipients of poor outcomes have limited redress when perpetrators adopt iniquitous procedures or tell outright lies. The 'crimes' of recipients are typically that they are better qualified, more attractive and popular, or hold political allegiances or opinions not appreciated by power holders. The strategies adopted by 'career bully' conduits, when repeated again and again, are likely to make even the most competent individuals with robust self-esteem feel exasperated and inadequate. Sometimes these conduits for *systemic* violence will boast of strategies adopted in previous 'career bully' appointments. For example, one perpetrator had his office painted black and a spotlight installed. When lower-level managers were called in, recipients were seated on a chair, the spotlight turned on, and the interrogation from the stalking 'career bully' began – and it worked – redundancies were grasped immediately!

It is always important to remember that 'career bully' conduits are hired to translate the wider economic or political pressures into a predetermined outcome for the organization. That is, the 'career bully' is the public face of the more covert underlying decision-maker (who may be a 'Lady MacBeth' conduit). As far as a firm is concerned, these behaviours are likely to work in the short term: staff will resign or accept redundancies in droves. Thus, 'career bullies' have many talents to offer to unscrupulous employers who need to downsize quickly when faced with a hostile environment. However there is a downside: if a career bully's lifetime behaviours continue, a firm is placed at both financial and legal risk from the *systemic* occupational violence, for example from vicarious liability or even criminal charges – in addition to productivity losses from other remaining staff who are negatively affected. Further, as McCarthy *et al* (2003a) have identified, bullying was a contributing factor to the persistence of corruption in a number of large international organizations that ultimately led to spectacular financial collapses, for example, at Enron, and in the London office of HIH Insurance. Thus, the persistence of 'career bully' behaviors after the strategies have served their immediate purpose can result in unanticipated negative consequences. What should a firm do with 'career bullies' once they have served their purpose?

The 'industrial psychopath'

An 'industrial psychopath' may also act as a conduit for *systemic* violence – but is particularly attracted to stressed organizations as these present an opportunity to act out sadistic fantasies with minimal personal risk. Babiak (1999) has argued that organizations that are undergoing re-structuring provide a fertile environment within which industrial psychopaths can flourish. That is, uncertainties rampant in an organization facing a severe and hostile external environment provide an ideal opportunity for an 'industrial psychopath' to advance. For example, the 'industrial psychopath' may lie about his/her own abilities or knowledge base to address a particular problem, steal core documents from other staff, report totally untrue 'facts' to the CEO to assassinate any opposition, exaggerate personal contacts with power holders to assist with difficult negotiations, disregard formal organizational policies, or sabotage or steal organizational property. The behaviours adopted by 'industrial psychopaths' are likely to fall within the criminal codes at times. Notably *malice* is *invariably* present as a motivating factor for these supervisory conduits of *systemic* violence. For example, those with sadist leanings may gain a perverse psycho-erotic thrill from exercising power over others in ways that cause them extreme discomfort. Unfortunately, CEO's may discover the harm to organizational goals and reputation too late.

Conclusion

It is clear that occupational violence is a predictable accompaniment to work in some jobs, and is in epidemic proportions in a few. In this chapter it was argued that there are a number of distinct types of occupational violence:

- 'External' violence perpetrated by an outsider to the organization, usually for instrumental reasons;
- 'Client-initiated' violence from a client or customer receiving a service;
- 'Internal' occupational violence/bullying from a co-worker or supervisor;
- 'Systemic' occupational violence which flows essentially from extreme economic pressures in the broader environment and which propels supervisors and colleagues into acting in inappropriate ways; and
- The risks of terrorism at work have been identified more recently for a range of workers in different industries across the world (see: Chapter 5).

However, there are multiple overlaps between the distinct forms of occupational violence/bullying. In particular, there is a significant overlap between 'internal' and 'systemic' occupational violence. The differentiating factor is that personal inadequacy is a core underlying feature common to perpetrators of 'internal' occupational violence, whereas economic and social forces external to the organization are the primary cause of *systemic* violence.

Historically, the workers at greatest risk of occupational violence were those who handled cash or valuables in the normal course of their work, and those with significant levels of face-to-face contact with clients/customers (Chappell and Di Martino, 2000). Because employment in most industrialized countries is increasingly skewed towards the services industries, an increase in the incidence and severity of 'client-initiated' occupational violence can be expected over the next decade. The evidence also indicates that 'internal violence/bullying' is now experienced more widely than is physical aggression in most workplaces. Further, increased globalization and internationalization of many market places can only add to the economic competition. Hence a dramatic rise in 'systemic' occupational violence may be inevitable.

It was argued in this chapter that *systemic* occupational violence only arises when two essential preconditions exist: (a) an excessively hostile external environment; and (b) an internal conduit person who accepts and then translates these excessive demands onto other workers.

(a) A *hostile economic environment* presents a number of challenges to CEOs and managers, including personnel skill and technique needs, product design and delivery changes, as well as service delivery alterations. Often restructuring of the organization will be needed, along with downsizing of the workforce, and changing production techniques. New ways of working can create a range of negative stresses for both workers and supervisors, particularly if organizational survival is on-the-line. In such an environment, uncomfortable decisions may have to be made. Whatever strategies are adopted (apart from firm closure), excessive and unreasonable workloads may result for some or all employees for variable periods of time. This fundamental point is often overlooked.

(b) *Internal conduits*: The second pre-requisite for the existence of *systemic* occupational violence is that an internal *conduit* person transmits these excessive and unreasonable demands onto other workers. While the behaviours adopted by supervisors and co-workers generally aim to enhance productivity or protect the survival of the organization, supervisory conduits may also adopt tactics that allow those with malevolent leanings to act inappropriately with impunity.

A typology of behaviors adopted by such supervisors (and others) as they translate systemic pressures onto workers was presented in this chapter. It was argued that there are ten basic types of behaviour that supervisors and co-workers adopt when faced with unremitting excessive pressure from the external environment. These 'ideal types' range from altruistic 'missionary' behaviour at one end of the 'continuum of malevolence' to the 'industrial psychopath' at the other end. Underlying each of the ten 'ideal types' of behaviour is the juxtaposition of individuals attempting to deal with unsustainable pressures imposed from outside.

Because many conduits are able to visualize the extent of the systemic pressures across an organization, they occupy a central position for cohesive articulation of resentment at the sources of the *systemic* violence. Unfortunately, it appears that this unifying rebellious role is rarely taken up. Rather, the over-riding mantra of organizational survival is widely accepted and incorporated without question. Where conduits consciously recognize that demands are unreasonable – and yet continue to transmit these excessive pressures on to others – they are complicit in the *systemic* process. Liefooghe and Davey (2001) have previously argued that extra-organizational environmental pressures facilitate internal bullying. It is asserted here that these inappropriate behaviours can be *legitimized* through a stated aim of organizational survival. Hence power relations are maintained, legitimized, and reproduced with both the perpetrator and recipients becoming the pawns dominated by the wider environmental pressures.

So how can this behaviour be averted? First, CEO's, supervisors and workers need to recognize where fault *really* lies, become cohesive, and attack the source. Second, a commitment to a 'zero tolerance of violence policy' needs to be widely adopted, accepted and enforced. The research evidence is overwhelming: most rational people make decisions after they have weighed up the risks. From both the disciplines of occupational health and safety and from criminology we know that the *certainty* of sanction has a greater deterrent effect than does the *severity* of the sanction (Zdenkowski, 2000; Scholz and Gray, 1990). Hence conduits of *systemic* violence will choose alternative behaviour patterns if sanctions inevitably follow inappropriate behaviour. Third, at a deep ethical level, individual workers, managers and supervisors need to internalize the belief that unremitting external pressure is not a valid excuse for people to behave inappropriately, abusively, violently or illegally. A few larger organizations have instituted expected 'codes of conduct' for staff, and these offer hope for more humane working lives for the wider workforce. Concomitantly, implementation of such 'codes of conduct' affirms managerial recognition that reasonable behaviours and systems of work contribute to overall productivity and well-being within organizations. Finally, because the pressures that lead to *systemic* violence are becoming endemic in western industrialized countries, more supervisors are likely to be tempted to externalize these pressures onto their subordinates. Economic cost/benefit calculations are needed that quantify the toll of human misery caused by excessive systemic pressures that develop into *systemic* occupational violence (see: Chapter 3).

In sum, the risks of 'external' occupational violence are well known, and preventive strategies have been widely promulgated by the police as well as other bodies. The risks of 'client-initiated' violence are becoming well-known, although because of the shift in the economic basis of society towards the services industries, an increasing incidence is probably inevitable. 'Internal' occupational violence/bullying (in its various manifestations) is also relatively common, although preventive measures are being implemented in most large organizations in industrialized countries. However, 'systemic' occupational violence is poorly understood at this point in time. Nevertheless, as the economic pressures on most organizations increase over time, supervisors are likely to increasingly translate these externally imposed pressures in a variety of coercive ways on subordinates. Hence appropriate control strategies are required to enable early identification of the potential for 'systemic' occupational violence so that preventive interventions can be implemented quickly.

The various typologies of occupational violence discussed above assist with understanding the distinct conditions that enhance, or diminish, the propensity for workplace aggression. These conceptualizations also assist organizations with preventive planning, and provide guidance on the variable behaviours that perpetrators may adopt under distinct scenarios.

The authors argue that effective safeguarding of an organization requires an integrated approach to risk management, including consideration of the broader economic environment (see: Chapter 3), community risk factors (see: Chapter 4), perpetrator characteristics (Chapter 5), employee and supervisor behavioural codes (see: Chapters 6, 8 and 10), worksite design (Chapter 4 and Appendix 1), and evaluation of preventive interventions (Chapter 10). Each of these aspects will be discussed in-depth in the following chapters.

3
Costs of Occupational Violence and Bullying

Paul McCarthy

The conceptual framing of costs and mediating factors that follows considers both tangible and intangible costs. Not all costs are reducible to financial terms, yet non-quantifiable costs can be significant. Non-quantifiable costs usually result from breaches of social, civil, political and cultural norms. Such norms can be expressed in indicators of best practice that provide a basis for accountability (see: Chapter 10). The conceptual framework proposed provides a basis for estimating the costs of bullying/violence. Given demonstrable cost, and the existence of preventive guidelines in most western countries, the question remains as to why incidences of bullying/violence remain relatively high in workplaces. For bullying/violence to persist in organizations permission must be given and benefit derived (Brodsky, 1976). Thus, the perverse counter case, that benefits accrue from bullying/violence at work, also needs consideration, in the interests of furthering accountability.

Cost of bullying at work

The estimates of costs that follow are presented on a per-case, organizational and national economic basis.

Per-case costs of bullying

Leymann (1990: 123) stated an early estimation that set out the parameters of the economic costs of a case of serious bullying/mobbing:

> We must assume that the economic consequences – like the psycho-social – are considerable. A person can be paid without having any work to do (or none at all), and this can go on for years. Long

periods of sick leave; a catastrophic drop in production by the whole group; the necessity for frequent intervention by personnel officers, personnel consultants, managers of various grades, occupational health staff, external consultants, the company's health care centres and so on... .All this extra effort, combined with loss of productive work, can be estimated to amount to between 30,000 and 100,000 U.S. dollars for the employees exposed to such mobbing.

A more detailed estimation of per-case costs of bullying in a public sector organization in Great Britain was stated by Hoel *et al* (2003: 156) as:

Absence	BP£6,972
Replacement costs	£7,500
Reduced productivity	*
Investigators' time for grievance investigation	£2,110
Local line-management time	£1,847
Head office personnel	£2,600
Corporate officers' time (including staff welfare)	£2,100
Costs of disciplinary process (hearing/solicitor)	£3,780
Witness interview costs	£1,200
Transfers	0
Litigation	*
Effects on those directly involved	*
Miscellaneous (effects on public relations, etc)	*
Total costs (minimum)	**£28,109**

(* unknown/difficult to assess) (At the time of writing $A1 was equal to 42.5 British pence)

Organizational costs of bullying

Financial costs of bullying are often annualised for comparative purposes. For example, Kivimäki *et al* (2000) estimated the annual cost of absenteeism caused by bullying in two Finnish hospitals at BP£125,000. This costing was based on findings that bullied employees had a 26% greater sickness absence than those not bullied over the prior 12 months.

Organizational costs of mobbing

In Europe, 'mobbing' (i.e. ganging up) has commonly been used to signify behaviours referred to as 'bullying' in the UK and Australia. Findings of surveys into bullying in the UK that identify a higher proportion of supervisors and less colleagues as perpetrators than do European studies of mobbing led Zapf (2001: 12, 13) to point to differences in definition, focus and usage between the terms.

The costs of mobbing in a German workplace employing 1,000 employees were estimated at US$112,000 per year (DM200,000), plus additional indirect costs of US$56,000 (Di Martino, 2000).

Costs of experiencing and witnessing bullying

Rayner's (2000) study in a UK workplace detailed costs of the impact on witnesses, including staff replacement, managers' and consultants' time, and costs of legal action. The costs were modelled for a 1,000 employee workplace in which 10% (i.e. 100) experienced victimization, leading to one in four victims (i.e. 25) leaving. One in five of the two witnesses to each case of bullying were also assumed to leave (i.e. 40). Replacement costs were estimated at BP£15,000 per employee, including '... generation of job specification, finding candidates ... preparing a shortlist, interviewing, selection, time lost because of low staffing levels (or temporary replacement cost), and the learning curve for the new incumbent' (Rayner, 2000:27). The total costs were:

One in four victims leave	£375,000
Two witnesses observe, and 20% leave	£600,000
Internal investigation	£75,000
Unfair dismissal action	£120,000
Total costs	£1,170,000

Rayner's estimation is valuable in drawing our attention to the costs of impacts on witnesses. The 10% prevalence ratio applied in Rayner's model is drawn from the conservative end of findings by Hoel and Cooper (2000) and UNISON (1997). An important caveat is that there may be less randomness and more self-selection amongst respondents in these studies than in random-sample studies completed by Leymann (1997). Also, differing uses of the cut-off for frequencies would have a significant bearing on calculations of costs. Assumptions about the relationship between bullying, intention to exit, and actual exit can also produce significant variations in cost estimates.

National costs of bullying

Hoel *et al* (2001) estimated the costs of bullying in Great Britain in 1999 within a range of £1.5–£2 billion (plus productivity loss and other unaccounted costs). The costs framework used by Hoel *et al* encompasses: effects on individuals (e.g. mental and physical health; and behavioural and attitudinal effects); organizational effects (including direct productivity and more intangible costs); and costs to society (e.g. retirement, medical care production loss and impacts on friends and family). A deductive approach was used in modelling the national costs of bullying. The following survey data and applications were used in the estimation.

- A nation-wide survey (Hoel and Cooper, 2000) that indicated that 10% of respondents had experienced bullying/mobbing in the last year, while 25% had been bullied in the last five years.
- The survey results that reported the 'currently bullied' group had taken 7 days more off work, on average, compared with those neither witnessing nor experiencing bullying. Also, absenteeism for those bullied in the past was at a higher rate than the 'not-bullied'. Overall, 15% of workers were considered affected by bullying. Absenteeism costs were estimated nationally at £1.5 billion in 1999.
- Replacement costs were calculated on the basis that 25% of those experiencing bullying would leave their jobs as a consequence, over the 3 years following the bullying (UNISON, 1997). Average replacement costs were set at £1,900 (see: Gordon and Risley, 1999). Total replacement costs were thus calculated at £380 million.
- Productivity and performance loss was estimated by applying findings from Hoel and Cooper's (2000) national-wide survey. The 'currently bullied' group indicated an average 7% productivity drop when compared with those that had neither experienced nor witnessed bullying. The group 'bullied in the last 5 years' indicated a comparable 4% drop in productivity. Overall, Hoel and his associates estimated that a 1.5–2% drop in productivity was possible. However, concerns about self-reporting dissuaded them from imputing a financial value to the impacts.
- Costs of internal complaints, grievances, retirement, and reputation were not estimated due to the lack of a reliable basis.

An integrated model of the costs of bullying

A comprehensive model of the costs of bullying (see: Table 3.1) was completed by Monika Henderson working with Paul McCarthy, Michael Sheehan, and Michelle Barker of the Workplace Bullying and Violence Project team at Griffith University (Sheehan *et al*, 2001). A range of annual prevalences was used, from 3.5% (see: Leymann, 1997)

Table 3.1   **Estimated cost per impact factor**

	3.5% prevalence (Au.$ million)	15% prevalence (Au.$ million)
Absenteeism among victims	235.7	1010.1
Staff turnover among victims	169.0	724.1
Absenteeism & staff turnover among co-workers	20.2–80.9	86.7–346.9
Legal costs for court & tribunal matters	44.9	192.4
Compensation costs for courts & tribunals	11.8	50.6
Compensation costs–conciliated/mediated	24.5–61.3	52.5–131.3
Redundancy & early retirement payouts	420.0	1800.0
Total overt direct costs	**926–1,024**	**3,916–4,255**
Formal grievance procedures	350.0	1500.0
Management/supervisor time addressing impacts	336.0–672.0	1440.0–2880.0
Workplace based support services eg EAP, HR	29.40–73.5	29.4–73.5
Workers Compensation costs	680.0	680.0
Total hidden direct costs	**1,395–1,776**	**3,649–5,134**
Productivity loss reduced performance by victims	390.4–1561.4	1673.0–6691.8
Productivity loss replacement employees	175.0–525.0	750.0–2250.0
Productivity loss internal transfer	17.5–43.8	75.0–187.5
Productivity loss co-worker	29.14–116.6	124.9–499.6
Productivity loss absenteeism	364.4	1561.6
Total lost productivity costs	**976–2,611**	**4184–11,190**
Intra-sector lost opportunity costs*	**2,609–7,828**	**5,219–15,656**
Out of sector flow on costs	**min. 35**	**min. 150**
OVERALL COSTS PER ANNUM	**5,942–13,273**	**17,118–36,384**

• *Increased to 1.0–3.0% for 15% prevalence rate estimate*

to 15% (see: Hoel *et al*, 2001). The frequencies of impacts of bullying were derived from a review of local and global research findings. The impacts were costed using Australian economic statistics. The model estimated costs of bullying at the following levels.

- National costs from Au.$6–13 billion (3.5% prevalence) including hidden and lost opportunity costs, rising to $17–36 billion dollars per year (15% prevalence) were calculated. The model also indicated that between 350,000 and 1.5 million Australian workers could be victims of bullying at work.

- Costs to smaller organizations (less than 20 employees) were between $17,000 and $24,000 per annum. Cost estimates for larger corporations that included direct, hidden and lost opportunity costs ranged from Au.$0.6 and $3.6 million per 1,000 employees per year.
- The average per-case costs of bullying were between Au.$16,977 (3.5% prevalence) and Au.$24,256 (15% prevalence).

In the model in Table 3.1, industry intra-sector lost opportunity costs were estimated in terms of:

- Failure of victims to take up training or promotion opportunities;
- Costs of inflexibility in work practices and hours;
- Undermining of creativity and innovation; and
- Degraded customer and client relationships and related market share, corporate image and shareholder impacts.

Estimates for out-of-sector flow-on costs encompassed:

- The business sector's share of public sector costs including public health and medical services following bullying (including prescription and counselling costs);
- Governmental benefits to unemployed victims;
- Costs to the public sector generated by formal complaints (e.g. administrative and legal); and
- Loss of profits resulting from the decline in spending by victims (and families) displaced from paid work.

The limiting of most of the costs in the model to the business sector is noted, together with the observation that they extend far wider, in the following terms.

Individual victims, their families, and co-workers experience social, psychological, and economic consequences as a direct result of workplace bullying. Neither have these impacts fully explored the hidden impacts of workplace bullying on the wider community. For example, the flow-on effects of bullying can affect both investors and the public. Such effects include the manufacture of inappropriate products, environmental impacts, poor decision-making on issues with wide-ranging effects, vulnerability to corrupt practices, poor customer service and the implications, for example, of public sector client needs not being considered or effectively addressed. (Sheehan *et al*, 2001: 9)

Costs of violence at work

There have been a series of studies that provide insights into the economic costs of overt occupational violence. Cost estimations are summarized below on a per-case and national economic basis.

Per-case costs of occupational violence

- The average cost of a single workplace homicide in the USA has been estimated at US$804,035 (Biddle and Hartley, 2002).

National costs of violence at work

- The British Crime Survey estimated that 3.3 million days were lost due to violence at work in 1997. The cost of these lost days was imputed at £180 million based on 'the compensation victims would have liked for the upset and inconvenience' (Budd, 2001).
- The British Columbia Worker's Compensation Board (Canada) reported that claims for loss of wages due to violent acts had increased by 88% since 1985 (ILO, 1996–2002).
- The cost of violence at work to employers in the USA in 1992 was estimated to be US$4.2 billion by the National Safe Workplace Institute (Di Martino, 2000). This figure can be compared with the Workplace Violence Research Institute's estimate of US$35.4 billion for 1995 (Kaufer and Mattman, 1998).
- The National Occupational Fatalities surveillance system in the USA reported that 14,000 workers were victims of workplace homicide between 1980 and 1997; an average of 800 fatalities per annum. Homicide was identified as the second leading cause of workplace fatalities, at an estimated cost of $12 billion to the US economy over this period (Biddle and Hartley, 2002).
- In New Zealand, loss of labour market income due to domestic violence has been estimated at NZ$1.2 billion in 1993/4 (ILO, 1996–2002).
- The costs of the costs of stress and violence at work have been estimated to be between 0.5% and 3.5% of GDP, and also to account for some 30% of all ill health and injuries (Hoel *et al*, 2001).
- An estimate of the costs of crime and crime prevention from a British Retail Consortium survey of 17,000 outlets totalled GB£2,044 billion in 2000 (Di Martino *et al*, 2003).

- We applied the cost of a single incidence of workplace homicide of US$804,035 reported in Biddle and Hartley (2002) to the World Trade Towers September 11 incidents. This method produced an estimate of the overall cost of workplace violence due to terrorism to the order of US$240bn in that year. Of course, the flow-on human, economic, social, political, cultural and military costs remain incalculable.

The estimation of costs of more overt forms of occupational violence are important in justifying investment in prevention. Wide variation in the use of cost factors and economic data are indicated in the reports outlined above. The development of more comprehensive models of the costs of occupational violence is clearly necessary. Cost factors could be drawn from the ILO (2003a) *Code of Practice*. There is also a need for surveys to collect more detailed information about the impacts of occupational violence to improve the basis of estimations. The logic of a generalized model to account for the costs of violence and bullying is outlined in the conclusion to this chapter.

Dimensions of cost

The 'costs' of bullying/violence include impacts that are subjective, intangible, individual and organizational are not amenable to financial expression. Nor are breaches of wider ethical, social, political, civil and cultural standards easily reducible to financial terms. The argument to consider both economic and non-economic dimensions of cost has been made by Hoel *et al* (2001). For example, fear and pain induced by incidents can also be considered as costs. Zero-tolerance of unacceptable behaviours is often more driven by ethical and political repugnance and fear felt widely in the community, than financial costs or statistical estimations of prevalence and impact. Such repugnance has been the driving force for regulatory constraints against forms of abuse/violence including domestic violence, child abuse, sexual harassment, stalking, and more recently bullying (McCarthy, 1999).

Our conceptual framing of the dimensions of bullying and violence and their possible cost impacts is presented in Table 3.2. The approach taken is that bullying and violence at work manifest in multiple forms and directions that are overlapping and interactive. The framework depicts impacts of bullying/violence at levels of the individual, the organization and the wider society (see: McCarthy, 1998; Hoel *et al*, 2001; Einarsen *et al*, 2003; Di Martino *et al*, 2003).

Table 3.2 Dimensions of the costs of bullying and violence

Forms of bullying/violence experienced at work	Individual costs/consequences	Organisational costs/consequences	Social, economic, political and cultural costs/consequences
Physical • body contact • threats • intimidation • negative body-language	*Health and well-being* • stress • anxiety • fear • depressive illness • psycho-somatic • heart attacks, strokes • PTSD • prolonged duress stress syndrome	*Productivity (recipients & witnesses)* • absenteeism, presenteeism • poor morale, commitment, loyalty • staff turnover • redundancy/early retirement • failure of quality, innovation • sabotage	*Social* • degrading of social capital • loss of trust • rise of anti-business & government sentiment • inter-group conflict • litigiousness
Verbal • swearing, yelling • denigration • humiliation	*Behavioural reactions* • avoidance • withdrawal • displacement • cynicism • learning • complicity • fightback	*Competitiveness* • poor quality product/service • loss of clients • poor reputation • staff recruitment difficulties • aversion of contractors, investors and other stakeholders	*Economic* • health, unemployment, welfare support • misallocation of resources • regional/national competitiveness • sustainability • regulatory, tax burdens
Social • back-stabbing • slander • exclusion • isolation • scapegoating		*Incident management* • managers, consultants time • EAP • rehabilitation, return to work • medical costs • legal costs, fines	*Civil* • regulatory failure • loss of transparency • corruption • poor public sector management • human rights infringements
Managerial/systemic • under-resourcing • work overload • impossible deadlines • denial of entitlements • aggressive administrative compliance • forcing resignation			

Table 3.2 Dimensions of the costs of bullying and violence – *continued*

Forms of bullying/violence experienced at work	Individual costs/consequences	Organisational costs/consequences	Social, economic, political and cultural costs/consequences
Direction • top-down • bottom-up • horizontal • to/from clients, contractors, strangers *Intention* • normalised in work culture • people skills failure • targeted • Machiavellian • revenge • pathological (socio) psycho-pathic, sadistic) • revenge • terror	*Work impacts* • stigmatisation • not promoted • transfer • reduced earnings • job loss • unemployment • ruin of career *Degraded relations* • work colleagues • family, friends • reputation • social status *Legal costs* • solicitors • barristers • expert opinion • grievances • tribunals • courts • appeals • awards of costs	*Insurance* • workers' compensation • property damage • public liability *Corporate threats* • corruption • targeting by regulators • suppression of information • financial loss, bankruptcy *Social accountability* • unethical work practices • poor ethical investment ratings	*Political* • apathy • alienation • resentment • disaffection • embarrassment • distortion of intelligence • terrorism *Cultural* • cultures of anxiety, fear, complaint, blaming, and aggression • loss of benefits of managing diversity • racial violence and bullying

Table 3.2 **Dimensions of the costs of bullying and violence** – *continued*

Forms of bullying/violence experienced at work	*Individual costs/consequences*	*Organisational costs/ consequences*	*Social, economic, political and cultural costs/consequences*
	Medical costs • consultations • medication • counselling • alternative therapies *Loss of income* • inability to work		

The potential for costs to escalate from the interplay of individual, organizational and regulatory factors is illustrated in the following case commentary.

'Fit in or fuck off' – a case illustration of cost escalation

(Note that the participants and events have been disguised, since legal action was undertaken).

An assistant manager who assumed family responsibilities became increasingly subject to negative reactions from his/her manager, including bullying behaviours, threats and intimidating body language. The assistant attributed the manager's reactions to the manager's belief that the recipient would no longer put in the long hours of overworking demanded due to the requirements of his/her family. As the bullying and intimidation continued, the assistant became fearful, agitated, and depressed. The assistant was also concerned about the manager's implication in financial impropriety and aggression to other staff members. Informal voicing of concerns by the assistant only served to intensify the attacks, and the recipient sought formal grievance proceedings. The manager then subjected the assistant to an aggressive performance appraisal, which was interpreted as a faultfinding exercise by the assistant.

The organization appointed another manager to investigate the grievance, however this new investigator was under the influence of the alleged perpetrator. While the investigation persisted, the assistant was suspended on full pay with nothing to do. The assistant's manager proceeded to make claims that his/her assistant's performance was unsatisfactory. The assistant then sought costly personal legal advice and his/her representations led to the appointment of a more independent investigator. However, this second investigator was unable to start the investigation for several weeks, due to prior commitments.

The assistant was next offered the opportunity to return to work in another part of the corporation, but considered the work meaningless and the offer humiliating. Mindful of the potential legal costs of action in industrial or civil courts, the assistant next pursued an application an anti-discrimination tribunal. This application was rejected due to difficulties in proving that the perpetrator's actions fell within the terms of the anti-discrimination statute. Next, an application was lodged with a crime and misconduct commission and the assistant then sought whistleblower protection, alleging that bullying/violence by the perpetrator was related to financial impropriety. Evidence of a prior history of bullying and violence by the alleged perpetrator was also presented.

At this point in time, the risks of escalating financial and other flow-on costs were considerable for all parties. Were the case details to find their way into the media, damage to reputational capital could be substantial – for the organization, officer bearers responsible for resolving the problem, and for the perpetrator. However, it is by no means inevitable that such scenarios will escalate. If possible interventions are widely known, and implemented, and escalation in severity of incidents can be avoided.

Factors mediating the translation of risks into incidents and costs

The presence of risk factors for workplace bullying and violence is accompanied by probabilities that incidents will occur, and costs will result. These probabilities can be reduced through mediating individual, organizational, and contextual factors (Axelby, 2000). Our model for the interplay of factors that precipitate costs of bullying/violence at work is presented in Figure 3.1.

The extent to which risk factors can reduce the severity and the costs of bullying/violence (per Figure 3.1) is proportionate to:

- *The recipient's awareness, resilience and coping capacities.* These capacities are influenced by: the recipient's understanding, personality, socialization, and prior experiences with conflicts and life pressures.
- *Risk assessment of the costs and benefits of pursuing different strategies in dealing with incidents* (see: Axelby, 2000: 45). An individual may trade the costs of pursuing remedies against potential benefits. Some recipients have a deeply felt sense of emotional hurt and injustice, overwhelming desire for reaffirmation (Rylance, 2002), revenge, and want financial compensation. In some cases, recipients may assume victim-identity and obsessively pursue remedies, regardless of costs (McCarthy, 2003). Organizations may tough-out claims for damages for vicarious liability in respect of failure of duty of care to protect their employees from bullying/violence. Often, the costs are displaced onto insurers who can profit from delaying actions for damages.
- *The quality of support and advice from a variety of actors*, including friends, work colleagues, family members, equity and harassment officers, HR managers, OHS representatives, employee assistance providers, union representatives, medical doctors, counsellors, and lawyers. Bullying and escalated risks of overt violence often result from the deterioration of conflict over time (Einarsen, 2000a). In such cases, the extent of costs is likely to be mediated by capacities of key actors to: encourage constructive conflict resolution; reconcile perceptions of bullying with definitional terms; accord natural justice to both the victim and the alleged perpetrator; distinguish vexatious complaints; and apply relevant organizational policies and procedures.
- *Cultural normalization of bullying and violence.* Some work and social cultures may normalize violent initiation rites, and dominant masculine cultures may neither regard certain behaviours as

Figure 3.1 Factors mediating risks, incidents, and costs of bullying and violence at work

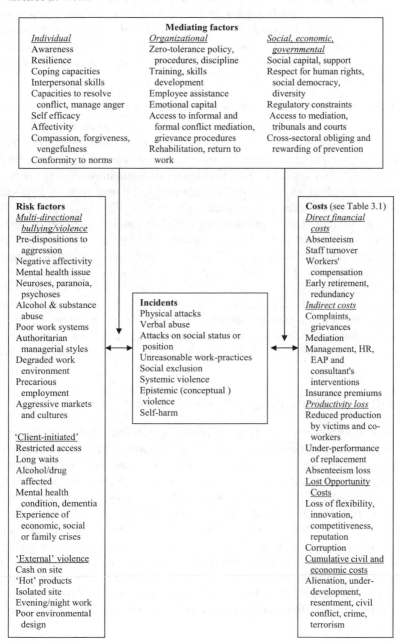

Mediating factors

Individual	*Organizational*	*Social, economic, governmental*
Awareness	Zero-tolerance policy,	Social capital, support
Resilience	procedures, discipline	Respect for human rights,
Coping capacities	Training, skills	social democracy,
Interpersonal skills	development	diversity
Capacities to resolve	Employee assistance	Regulatory constraints
conflict, manage anger	Emotional capital	Access to mediation,
Self efficacy	Access to informal and	tribunals and courts
Affectivity	formal conflict mediation,	Cross-sectoral obliging and
Compassion, forgiveness,	grievance procedures	rewarding of prevention
vengefulness	Rehabilitation, return to	
Conformity to norms	work	

Risk factors
Multi-directional bullying/violence
Pre-dispositions to aggression
Negative affectivity
Mental health issue
Neuroses, paranoia, psychoses
Alcohol & substance abuse
Poor work systems
Authoritarian managerial styles
Degraded work environment
Precarious employment
Aggressive markets and cultures

'Client-initiated'
Restricted access
Long waits
Alcohol/drug affected
Mental health condition, dementia
Experience of economic, social or family crises

'External' violence
Cash on site
'Hot' products
Isolated site
Evening/night work
Poor environmental design

Incidents
Physical attacks
Verbal abuse
Attacks on social status or position
Unreasonable work-practices
Social exclusion
Systemic violence
Epistemic (conceptual) violence
Self-harm

Costs (see Table 3.1)
Direct financial costs
Absenteeism
Staff turnover
Workers' compensation
Early retirement, redundancy
Indirect costs
Complaints, grievances
Mediation
Management, HR, EAP and consultant's interventions
Insurance premiums
Productivity loss
Reduced production by victims and co-workers
Under-performance of replacement
Absenteeism loss
Lost Opportunity Costs
Loss of flexibility, innovation, competitiveness, reputation
Corruption
Cumulative civil and economic costs
Alienation, under-development, resentment, civil conflict, crime, terrorism

bullying/violence, nor recognize unacceptable costs to individuals, organizations or the wider society (Ehrenreich, 2001). Constraints to bullying, in particular violence, are more likely to be found in feminized cultures and workplaces (Einarsen, 2000a; McCarthy, 2001). On the other hand, feminized workplaces may be more vulnerable to external violence, particularly in service industries.

Differences in costs of bullying and violence over time

Definitions of occupation violence have, at least until recently, emphasized overt incidents that have the potential to manifest physical and/or psychological injury from just one single event (Di Martino *et al*, 2003: 3). Costs of violent incidents can spiral immediately following the incident, escalate, or plateau and decline over time as physical and/or psychological injuries heal.

It is notable that definitions of bullying emphasize repeated less overt negative actions. The cumulative effects of bullying act to degrade the recipient's psychological integrity over time. The longer the bullying is experienced, the more extensive and enduring is the psychological injury (Leymann, 1996; Einarsen *et al*, 2003). Thus, bullying has the potential to manifest psychological trauma to at least the same extent as more overt violence (Leymann and Gustafsson, 1996). Some physical injuries may heal more quickly than the progressive desiccation of a victim's psyche caused by the experience of repeated bullying.

Issues in costing bullying and violence at work

Data used to cost bullying/violence are derived from surveys into prevalence and impacts. Self-selected samples are likely to give higher prevalences, and lead to higher estimates of costs than random sampling. Surveys to estimate prevalence usually ask respondents about the extent to which they have experienced: (a) bullying/violence, as they understand the concepts; (b) behaviours that fall within a definition provided; and (c) specified negative acts or a combination of such questions. The rigour in the definitions, extent of examples given to respondents, and the manner in which cut-offs are applied to experiences of negative acts, all influence prevalence ratios. In the case of bullying, the approach described in (a) above can produce higher frequencies of bullying than (b) or (c) (see: Zapf *et al*, 2003).

Costs of failures of natural justice and due process in resolving complaints of bullying – a case commentary

(Note that the events have been disguised, since the case was subject to legal action)

The denial of natural justice and due process accorded to both victims and perpetrators of bullying/violence at work can contribute to costs of absenteeism, staff turnover, and lost productivity. Such costs can also be identified as departures from standards of good practice recommended by governments, regulators and professional bodies. Such risks may be difficult to quantify and of low probability, however they can be of extreme consequence.

As an illustration of these dynamics in practice, consider a case in which a recipient alleged a manager had deliberately jostled him/her as s/he passed them in a hallway. The entitlements of both the recipient and the perpetrator to natural justice and due process include both having the right to present their side of the story together with any evidence from witnesses. In this particular incident there were no witnesses, however, the recipient alleged that the incident was part of ongoing negative behaviours from the perpetrator. The alleged perpetrator responded that s/he had accidentally bumped the complainant in negotiating a turn in the passageway. The parties in such a dispute are also entitled to have their evidence adjudicated in a balanced manner by a competent person. This was not to eventuate in the case described. The matter was not well handled by senior managers who did not have experience in resolving such complaints, and were also associates of the alleged perpetrator.

In the case described above, prospects for an early resolution of the matter could have been enhanced through independent mediation in a no-blame, non-stigmatising manner. Developmental goals could also be sensibly agreed, if warranted. However, there was no such resolution and the alleged perpetrator retaliated by negatively appraising the performance of the recipient, effectively 'blaming the victim'. Where the negative actions experienced by the victim can be located within a series of negative actions of which there is some evidence, then a more escalated, disciplinary response could be justified. In this case, the recipient later reported an incident for which there were witnesses, alleging the perpetrator had manipulated an office fitting that knocked the complaint to the floor. The perpetrator's defence was that s/he did not see the complainant when s/he commenced moving the fitting. In response, the recipient claimed that the design of the fitting enabled visibility of a person standing on the other side.

Over time, the failure to accord natural justice, due process and appropriate resolution of these and other incidents led the recipient to suffer psychological injury and to leave the company. Subsequently, the recipient took civil legal action against the company claiming vicarious liability and considerable damages for failure to maintain a safe place of work. The case reached the court some four years after the recipient experienced his/her first incident of bullying/violence from the perpetrator. Success in the recipient's action would have bankrupted the company, and the parties agreed to settle the claim out of court for an undisclosed sum.

For a recipient to qualify as having been bullied/mobbed in Leymann's LIPT Questionnaire (1997) they would have to experience at least one of a list of some 50 negative actions, once a week, for at least 6 months. This definitional cut off was consistent with the notion that the severity of the impacts of bullying/violence accumulated over time. Leymann (1997) used this cut-off in a study generalizable for the entire Swedish workforce and found a 3.5% prevalence of mobbing for a 6-month period.

Higher prevalence rates often result from studies that use less restrictive definitions than Leymann's. Caution needs to be exercised in the application of cut-offs in the estimation of the extent to which trauma is experienced and costs. Commonly, survey respondents indicate a range of impacts as a consequence of varying degrees of exposure to bullying/violence over time. The impacts, prevalences and frequencies of the experience of bullying/violence are averaged across sample populations and the findings projected across organizational, regional or national populations to provide a basis for the application of economic statistics in modelling costs. The extent of impacts varies widely across survey studies (see: Sheehan *et al*, 2001). Variation in the basis of costing across nationalities also makes international comparisons difficult (Hoel *et al*, 2001).

Notwithstanding these limitations, survey-based approaches are likely to produce more useful estimates of cost than either case reports, organizational data bases, or records maintained by governmental agencies. Only 10% of actual incidences of bullying and violence at work are likely to be reported (Chappell and Di Martino, 2000). Likewise, official crime, health and safety and worker's compensation data are under-reported (Mayhew, 2003). Both organizational and official records can be biased to injuries produced by more overt violence. The link between assault and injury may also be more clearly causal than that between bullying and negative health, attitudinal or behavioural effects (Hoel *et al*, 2001). Also, per-case costs that accumulate over a number of years may be difficult to annualize.

The transfer of costs between groups is another reason for caution. For example, a national economic cost, such as social security, may be an income to an individual (Hoel *et al*, 2001). Organizations may also displace costs of serious incidents to insurers. Flow on costs from organizations to the wider society can lead to individuals and corporations paying higher levels of tax. The next section concentrates on the costs of bullying, followed by and analysis of more overt occupational violence.

Benefits of bullying/violence at work – the perverse counter case

Reasons why workplace bullying/violence persist in organizations in the face of evidence of the costs are the focus of the discussion in this section. Recipients routinely trade-off the costs and benefits of remaining in situations where they are exposed to bullying/violence. This reality is perhaps expressed in the under-reporting of bullying and violence (Chappell and Di Martino, 2000). A recipient may defer to a perpetrator out of lack of awareness that their rights are being infringed. They may also defer out of fear, in the interests of keeping the peace, their job, and careers – Machiavelli's advice was to stay close to your enemies.

Some employees can react to bullying by working harder, as though to demonstrate commitment and performance in the face of the threat posed to their self-esteem and position in the organization (Hoel *et al*, 2001: 32). Identity-formation can sometimes involve taking victim-identities that are unhealthy or deathly, for example, in the negative self-affirmation of the neurotic, anorexic, or masochist (McCarthy, 2001). Consumption practices pursued in identity-formation can also normalize unacceptable work practices (McCarthy, 2003a).

An historical lineage of 'organising-violence' in the extraction of profits from human labour has been identified (McCarthy *et al*, 1995). Notwithstanding the voluminous managerial literature and corporate PR about participation, empowerment, and teamwork, the evidence that people-centred management improves performance over more authoritarian styles remains sparse (Ironside and Seifert, 2003: 385). Management styles that are authoritarian, even fascist, are more common in history than participative, democratic styles (see: Mailer, 2003). Short-term benefits may accrue to personalities who perpetrate bullying styles of management as the 'way things are done around here'. Much of Al Dunlap's (1996) *Mean Business* reads as tough management and work organization that strays across the line of reasonable management practice into bullying and systemic violence.

In recent years, threats of illegal immigration and terrorism have contributed to the re-legitimation of tough management targeted against the enemy from within and without. For example, the violent style of managing the arrival, detention, and processing of refugees who arrive in Australia is argued, by the government of the day, to confer benefits that outweigh the costs.

The evolution of civil society and social democracy, coupled with legislation to protect employees continues to precipitate ethical-political debate about the crossover from appropriate to inappropriate levels of coercion/violence. Such debate has produced guidelines and regulations that endeavour to internalize the costs of hitherto normalized bullying/violence that have otherwise been externalized to individuals and the wider society.

Debates about costs and benefits around the fine line between aggressive performance management and bullying bring assumptions about the quality of the work environment, regulatory and professional standards, market exigencies, human nature, and rights into play. Such debates are conducted in terms of ethics, responsibility and politics that transcend questions of economic cost. These debates are positive in addressing concerns about the quality of working life and reasonableness of management practice in organizations impacted by pressures of global market capitalism. The debates are also positive in scrutinizing costs of bullying/violence implicated in poor corporate governance and failures of social responsibility (McCarthy *et al*, 2003a).

Conclusion: accounting for costs of bullying/violence at work

Most corporations do not presently account for the costs of bullying and occupational violence in a comprehensive manner. Furthermore, bullying and violence at work still remains a relatively undiscussed topic in workplaces (Einarsen, 2000b). In these circumstances, bullying/violence persists and generates costs due to lack of awareness, fear of the consequences of complaining/reporting, and lack effective preventive measures. Formidable costs for complainants include the risk of stigmatization, further negative reactions from perpetrators, loss of earnings and career opportunities, and the indeterminate cost and duration of remedies.

Concerns with accounting for the costs of bullying and violence at work can be located in a lineage of Human Resource Information System (HRIS) initiatives to account for the quality of the human asset and of emotional capital and social capital at work (McCarthy, 2003b). Findings of research funded by the Swedish Work Environment Fund in the 1980s and 1990s prompted Leymann's (1990: 123) early estimation of the costs of bullying. The Swedish government's interest in the costs of bullying reflected concerns shared with business and unions about productivity and wider social costs of bullying at work. Leymann's research validated concerns that failure to prevent bullying resulted in degraded productivity and sick leave and resignations. The potential costs of interventions, loss of skills, and burden on retirement funds led to the implement of the Swedish code *Victimisation at Work* (1993). Concerns about costs remain particularly relevant today, and calculation of the costs of bullying and violence is necessary to justify investment in prevention by organizations, industry bodies and governments.

There is a case for case for organizations to calculate the *probable actual costs* of bullying/violence in their HRIS based on valid industry studies. Greater awareness of costs within an organization can be gained by adopting accounting procedures for costing bullying/violence. The following approach is recommended.

1. The *probable actual costs* in the organization (including likely unreported incidents) can be estimated using available survey findings indicating prevalences and impacts relevant to the industry sector. These prevalences and impacts can be converted to financial costs for staffing levels in the organization using national statistical data. In the absence of valid data gained through random-sample surveys, Leymann's (1997) finding of a 3.5% prevalence remains a useful base-line indicator for estimating the costs of bullying. Hoel *et al* (2001) estimated that 6% of European Employees experienced physical violence per year (2% from fellow employees and 4% from outsiders) and that 10% were subjected to bullying.
2. Costs that are difficult to quantify can be reported as departures from best-practice standards for ethics, sustainability, governance, human rights and the statutory regulations. Examples of such risks include bullying/violence that enables corruption to persist to the point of corporate collapse (McCarthy *et al*, 2003a), and risks of violent homicidal reactions including massacre.
3. The organizational HRIS can be extended to record events and costs of bullying and violence collected from incident reports and exit interviews. These reports provide a basis for recording *the costs of reported incidents* in an organization.
4. A provision for *probable actual costs* can be made in the HRIS as the basis for comparison with actual recorded costs. This comparison could be undertaken on a two-year cycle, and include evaluation of the organization's preventive policies and procedures. Where progressive survey and evaluation findings indicate that a preventive policy is working well, provisioning for *probable actual costs* can be reduced. The comparison can also provide a basis for justifying new preventive interventions.

5. The organization's triple bottom line accounting system can report on measures to prevent bullying/violence in terms of indicators of best-practice in sustainability and social accountability rating systems (McCarthy *et al*, 2003).

Such a comprehensive approach would also re-internalize costs of bullying/violence that have been externalized to individual employees and the wider society, whether through meanness, short-termism, or management failures within organizations.

4
The Overlap between Community Violence, Criminal Indices and Aggression at Work

Claire Mayhew

The discussions in this chapter are focused on identification of country, geographical area, and community/cultural risk factors that may spill over and manifest as one or other form of occupational violence. It is argued that cultural norms of behaviour (including attitudes to violence) and external pressures permeate the factory/office boundaries. That is, organizations are embedded within the social and cultural frameworks of their environment. Hence if the strength of precursor risk factors for violence/bullying varies significantly across geographical areas *within* countries, the incidence of occupational violence may also alter between organizations based in different suburbs. The author of this chapter hypothesizes that the risk of 'spillover' occurring is predictable, providing reliable precursor data are available and are objectively assessed.

Data on violence

The data that may help predict the risk of 'spillover' between community violence and the workplace include:

- Homicide ratios in different countries;
- Homicide ratios in different jurisdictions/geographical areas within countries;
- The level of concentration of people with characteristics commonly associated with violence in particular areas;
- Early discovery of emerging forms of aggression; and
- The extent to which effective preventive interventions are disseminated and implemented.

The underlying hypothesis is that aggression has multiple causes, may be a *learned* behaviour in different contexts, and may be acted out in several environments (Graycar, 2003). For example, a young male who has experienced violence at home and bullying at school may react aggressively in any social or work context where he feels threatened, whether from status threat, financial stress, desperation, rejection, or depression about future prospects. The author believes that violence directed to the owner/managers of small retail businesses may be an early indicator of the level of general aggression and desperation in localized communities. Hence the final part of this chapter examines the range of crimes that may be directed to retail small business owner/managers.

At the outset, it is useful to review what aggressive behaviours are legally defined as 'crimes' in most industrialized countries (and hence come within the criminal codes) and those that do not.

- First, some violent behaviours have been in criminal codes for years across the industrialized world. These crimes are usually feared by citizens and receive significant media attention, such as robberies with violence. The 'popularity' of different types of these crimes changes over time as opportunities vary. For example, there appears to be a recent surge in 'popularity' of hold-ups of individuals using ATM's.
- Second, there are non-violent crimes that impact on the financial bottom line of individuals and businesses and which many people fear can lead to violence or significant deprivation, such as the 'break and enter' of houses, fraudulent use of credit cards, or scams involving significant amounts of money. While violence is rarely premeditated during the planning of such offences, physical force may be used by perpetrators when making attempts to escape; or offenders may become overtly aggressive to bolster his/her denials when apprehended.
- Third, there are behaviours that frighten and intimidate individuals but which do not constitute 'crime' in the legal sense of the word. For example, groups of young people may congregate around retail shops and discourage potential customers from entering and spending their money.

As a backdrop to identification of geographical area variations in levels of risk, some differences in the levels of violence experienced in diverse countries are identified.

Variations in experiences of violent crime across countries

There is a range of ways in which the extent of victimization can be measured within and between countries. The overall international pattern indicates that violent crime has risen by about 22% over the past 4 years (Barclay and Tavares, 2003: 4). There are two fundamental ways by which victimization can be assessed: (a) through criminal justice system data, such as convictions for violence; and (b) via surveys of representative samples of the population.

Criminal justice system data: an international overview

The criminal justice system within each industrialized country records information about a range of offences and *convicted* perpetrators. These data can usually be separated out on the basis of perpetrator and victim characteristics, victimization rate per 100,000 population, and (sometimes) percentage changes over time. Unfortunately, different countries define offences in distinct ways because of varying legal definitions, adopt various statistical data collection processes, and enact diverse rules governing classification of events. Hence international comparisons can be very difficult.

In order to enhance comparability of patterns of victimization across countries, the International Crime Victim Survey (ICVS) was initiated in 1989, followed by later sweeps in 1992 and 1996/97, with the fourth ICVS analysis in progress at the time of writing this chapter. There is a broad range of crime patterns identified and analysed within the ICVS, only some of which are relevant to the focus of this book. Notably, the incidence of homicide, physical assault and sexual assault provide useful indicators of proclivities to violence within countries, and also allow some basic comparisons to be made between nation states.

The most unambiguous identifier of violence in a country is homicide because international definitions are similar. Hence reliable international comparisons are able to be made. In the following table, information about homicides in selected countries is shown, based on the average homicide rate per year per 100,000 persons over the period 1998 to 2000. As can be seen, there are significant variations in the level of risk of homicide between nation states.

The data shown in Table 4.1 indicate that South Africa and Russia have a notably higher incidence of homicide than other country populations covered in this evaluation. The Baltic States and USA also have a higher incidence than the member states of the European Union, Canada, Australia and New Zealand.

Table 4.1   **Homicides per 100,000 population for selected countries: average per year 1998 to 2000**

	Homicides per 100,000
Australia	1.87*
EU member states average	1.70
England and Wales	1.50
Austria	0.90
Belgium	1.79
Canada	1.79*
Denmark	1.00
Estonia	11.43
Finland	2.60
France	1.68
Germany	1.19
Greece	1.55
Japan	1.06
Latvia	6.51
New Zealand	2.28*
Norway	0.92
Russia	20.52
Spain	2.77
South Africa	54.25
Sweden	2.06
Turkey	2.54
United States	5.87

(Extracted from Barclay and Tavares, 2002: 10; based on the 2000 ICVS analysis)
* A different study with an adjusted definition identified the same homicide rate of 1.8 per 100,000 in Australia, Canada and New Zealand for the year 2000 (Segessenmann, 2002).

Another indicator of the level of violence endemic within a country is the risk of victimization through non-fatal violent events. However, violence that does not result in a homicide is more likely to be defined and classified differently in separate nation states, and to go unreported to the police. Hence the reliability of international comparisons is lower than for homicides.

The data in Table 4.2 indicate that although there are some similarities in the patterns of violent crime data reported across industrialized countries, there are also some significant differences. Some of these differences between countries in reported violent offences are due to different definitional and recording protocols:

> ...the USA and Canada do not appear to include minor assaults, intimidation, and threats within their definition of violent crime.

However, New Zealand does include these crimes in its definition, and these offences comprise approximately half of all violent crime in this country. Also, New Zealand does not include sexual offences in violent crime, whereas Australia, USA, Canada, England and Wales do (Segessenmann, 2002: 3).

Table 4.2 Victimization risk: percentage of population victimized by a 'contact crime' on at least one occasion*

	% Victimized
Australia	4.1
England and Wales	3.6
Belgium	1.8
Canada	3.4
Denmark	2.3
Finland	3.2
France	2.2
Japan	0.4
Spain	1.5
Sweden	2.2
United States	1.9

(*Based on 1999 recorded data on robbery, assaults with force, and sexual assaults against women; extracted from Barclay and Tavares, 2002: 17.)

That is, apart from homicides, the international comparative data on violent crimes must always be interpreted with caution. The available international data also indicate that the risks of violence are altering over time, but to a different extent in various countries.

Violent crimes recorded by the police *increased* by 72% in Japan, 49% in Poland, 38% in Spain, 36% in France, 24% in Australia, 23% in Lithuania, 23% in South Africa, 15% in England and Wales, 14% across the European Union, 6% in New Zealand, and 2% in Canada over the period 1996 to 2000 (Barclay and Tavares, 2002: 12). In contrast, violent crime *decreased* by 49% in Eire, 37% in Cyprus, and 16% in the US (Barclay and Tavares, 2002: 12).

In *Australia* – as in many other countries – the tendency has been to rely on criminal justice system data rather than population surveys, particularly for *occupational* violence indices. Sole reliance on reported criminal justice system data has a number of weaknesses. For example, events not reported to the police, and those where criminal convictions do not eventuate may remain outside of formal data bases. These deficiencies can be overcome (to some extent) by population surveys.

Surveys of the population

Victimization ratios can be estimated through regular surveys of the population who are requested to provide confidential information on selected experiences. Such surveys allow improved understandings of the *contexts* when violent crimes most commonly occur, the impacts on the recipients, and may also identify a range of events *not reported* to the criminal justice system (see: Morris *et al*, 2003).

As one example, the *British Crime Survey* is conducted at regular intervals over large nationally representative household populations and the de-identified findings are widely published. For example, the detailed report *Violence at Work: Findings from the British Crime Survey* provides information about the extent and nature of occupational violence in England and Wales (published jointly by the Home Office and the Health and Safety Executive). In 2000, the *British Crime Survey* data indicated that there were almost 1.3 million incidents of occupational violence (including 634,000 assaults); the risks varied significantly across occupations with 'security and protective service' workers (11.4%) and nurses (5%) at greatest risk of an assault on-the-job (Budd, 2001: 20). Similarly in Australia, the risks vary with up to 80% of disability care workers assaulted in a year (Ore, 2003: 232). Of concern, Budd (2001: 6) reports that: '... incidents of violence at work increased by 5% between 1997 and 1999...assaults increased by 21%, while the number of threats fell by 7%'. The British Health and Safety Executive was sufficiently concerned by the incidence – particularly the *increasing incidence* – that it launched a 3-year programme to reduce the incidence of occupational violence by 10%, including in the National Health Service and the London Underground (who experienced 35% of all threats recorded in the *British Crime Survey)* (Harris and Wilson, 2002: 16).

The significant variations between countries (such as shown in Table 4.1) suggest that local risk factors, diverse social pressures, culture and/or regulatory controls may influence the risk of violent victimization. For example, in New Zealand the increased risks for Maori residents have been identified through surveys; including an increased tendency to non-reporting among this ethnic group (Morris *et al*, 2003). It is logical, therefore, to assume that even *within* countries the risk of experiencing violence may vary from place to place as precursor factors diverge across geographical areas and/or suburbs.

Variations in risk of violence/bullying between geographical areas and communities

Based on *US* violent and property crime data, it has been argued that two sets of societal characteristics influence the level of crime: (a) the degree of relative deprivation (homicides and assaults are strongly associated with relative deprivation and low social capital); and (b) the level of social cohesiveness between citizens (Kawachi *et al*, 1999). In addition, the variable ability to access firearms continues to be a core risk factor associated with the *severity* of outcomes. That is, the comparatively easy access to firearms in the United States may be a core explanation why – compared with other industrialized countries – the homicide risk is so high in that country (see: Table 4.1). Indeed, as this chapter was being finalized at the end of September 2003, there were media reports of a 4-year old child in the USA who 'played' with a gun – with fatal consequences. Similarly, Ananthaswamy (2003) cites evidence that people who keep guns at home have a 72% greater chance of being killed via a gun than those who do not.

If tendencies to violence in the general community are not homogenously distributed across all members of a population, specific groups of workers in particular geographical areas may be at significantly increased risk. There is a multiplicity of potential reasons, including:

> ... members of an organisation located in a violent community may learn aggressive behaviour through observational learning, symbolic modelling, and imitation of the violence in that community (Dietz *et al*, 2003: 318).

Dietz *et al* (2003) studied 250 geographically dispersed public sector work sites (each with an average of 680 employees), longitudinal data from different sources, the violent crime data for three years (murders, manslaughters, rapes, robberies and assaults etc), worksite data, and information from a random sub-sample of 5,419 survey respondents. During analysis the researchers controlled for variables that have traditionally been reported to be predictors of violence, including economic deprivation and family disruption. While initially Dietz *et al* (2003) hypothesized that the internal culture of an organization could safeguard the organization from aggressive behaviour by workers through enhancement of the *procedural justice climate,* the researchers concluded that:

> ...community-level violent crime rate was positively associated with the workplace aggression count, whereas the procedural justice climate had a marginally significant, negative zero-order relationship with the workplace aggression count (Dietz *et al*, 2003: 321).

That is, violence in the community surrounding a worksite spilled over and predicted 'internal' occupational violence – and the organization's procedural justice climate did *not*. The authors concluded that workplace violence was a complex phenomenon with multiple causes which spilled over from the community, reflecting role modelling, learning, and normative influences on workers (Dietz *et al*, 2003: 323).

Similarly, based on international data from several countries, Di Martino (2002) argued that occupational violence could occur as a 'spill over' from local communities, resulting in an acute risk for particular groups of workers:

> Several country surveys strongly hint at direct inroads of societal problems into the workplace with violence as one of the major issues. This exposure and permeability have always been known to be important factors in the triggering of workplace violence. It is now shown to be of extreme relevance as far as the health sector is concerned. The sector appears, by the very nature of the activities that are carried out therein, to be particularly exposed to external influences and as such, especially vulnerable to violence from the outside (Di Martino, 2002: 24).

More detailed information about the violence risks faced by health workers are presented in Chapter 6. For all industry sectors, there are a range of possible reasons why the risk of violence at work is concentrated in particular geographical areas and not others.

Changes in employment opportunities, competition, and unmet social support needs

As a result of changing employment and demographic patterns, the precursor risk factors for violence and other criminal activities may be concentrated in particular geographical areas. Where unemployment (or under-employment) is higher, rapid social change is occurring, or a greater proportion of the population experiences relative deprivation, increased frustration may spill over into aggressive acts in a range of environments.

In geographical areas where there is a rapidly expanding population, inadequate infrastructure may compound stresses, for example, poor public transport, a lack of municipal housing, insufficient educational facilities, or stretched emergency social support services. Thus, when employment opportunities or demographics rapidly change in a local area (for whatever reasons), the demand for a range of social support and other services may outstrip the supply.

Globalization and enhanced competitiveness have also had major repercussions on working populations across the industrialized world. There has been a wave of downsizing across larger private enterprise and public sector organizations, with a corresponding increase in employment in small-scale businesses, contract, casual and short-term employment arrangements. When these employment shifts occur rapidly, individuals may feel threatened, resentful or frightened of the repercussions on their lifestyle and income. In the US, in particular, there have been frequent reports of mass shootings perpetrated by employees following unwanted redundancies and/or poor performance appraisals. The skills of some workers may also become obsolete rapidly as technological innovations are developed and implemented. Those who are unskilled or semi-skilled, unable to alter their skill sets to the 'new' requirements, who are held back from up-skilling through physical/mental health restrictions or familial or social responsibilities, may be locked out of the labour market. When this happens in remote areas, unemployed workers may have to rely on social security for minimal income, or relocate to larger centres away from home base.

Remote communities

While *remote* geographical area data are poor, there are clear indications that those living in isolated communities experience a far higher incidence of domestic and other forms of violence; a risk that is exacerbated by the increased availability of firearms (Maskell-Knight, 2003: 27). For example in Australia, the New South Wales Bureau of Crime Statistics and Research has collated data on prosecutions for violent offending, separated by Local Government Areas (LGA). Over the 12-month period January to December 2003, the incidence of apprehended and prosecuted assault offenders in LGA communities was, proportionately, higher in rural and remote areas of that state with 23 of the 25 top-ranked LGA's for assault outside metropolitan areas (BCS&R, 2003a). Notably, most of these were also lower socio-economic localities (ibid). The incidence of sexual assault showed a similar distribution (BCS&R, 2003b).

The scholarship of Williams *et al* (2003) highlighted the distinct risks of violence faced by Indigenous workers vis-à-vis other employees in Australia and which are clearly linked with community strain (proportionately more Indigenous Australians live in 'outback' areas). For example, she reports that 62% of Indigenous health workers interviewed cited abuse or threats from client/patients over the previous 12-months. The risk factor environments varied:

> ...mostly are drug addicts ...from clients if they are not picked up on time ... screamed at ... relatives and friends escorting the patients from the Aboriginal community... (Williams *et al*, 2003: 57).

The risk is exacerbated in rural and remote areas if there is limited emergency back-up. In all these scenarios, nascent aggressive tendencies in stressed individuals can be fuelled by financial or social pressure.

Individual risk factors

A consistent finding across criminal justice system data is that perpetrators of violence are disproportionately young males. If young people are excluded from employment, addicted, or are unable to survive economically they may be tempted to commit a range of crimes, including violence. Among this 'at risk' group may be a cohort of angry young men who live for the moment, have few future prospects, and who seek gratification when and where they can obtain it (Graycar, 2003; Graycar, 2001). For example, the New South Wales Bureau of Crime Statistics and Research reported that 83% of all apprehended and prosecuted offenders were male, with the bulk in the young adult range (BCS&R, 2003c). Assaults and sexual assaults were disproportionately committed by males aged 18 to 19, followed by those aged 20 to 24 (BCS&R, 2000).

Recipients of violence and homicidal attacks share similar demographic features to the bulk of perpetrators. The Australian Institute of Health and Welfare primary data for the Australian population as a whole indicate: (a) young adults are at significantly increased risk of violent victimization; and (b) facial fractures, intracranial injuries and open wounds are very common consequences (Stevenson, 2001). Similar findings have been reported in New Zealand (Morris *et al*, 2003).

A recent 'innovative' and potentially violent community-based behaviour has been titled 'swarming'. Mass 'swarming' – sometimes with violent rioting – occurs when hundreds of uninvited young people gatecrash parties at homes after being alerted via mobile phone text communications (Toohey, 2003). Theorists have linked this new phenomenon with the extended dependency of adult children, unemployment or underemployment, lack of adult responsibilities, reduced home security compared with hotels and public places, social status compensation behaviour from marginal middle-class young men, the natural progression of de-sensitisation, the staking out of territory from other teenagers and police, and ritual confrontation that feeds off itself – all of which have resulted in a trial curfew for young people in Western Australia (Toohey, 2003: 19).

Alcohol and illicit substance use

It is likely that the stresses that accompany unemployment/underemployment and rapid social change may also draw people to substance abuse, which in turn will increase the need for additional income. Hence it is unsurprising that the international research literature also shows a consistent link between intoxication and violence. In particular, younger males intoxicated with alcohol or illicit substances are consistently over-represented among convicted violence perpetrators (BCS&R, 2003c; Makkai and McGregor, 2003; Muscat and Carcach, 2002). Homel (1997) conducted a series of studies of the relationship between alcohol and violent offending around Australia's Gold Coast nightclubs and elsewhere in this locality. The findings consistently showed that young males affected by illicit drugs or alcohol were higher-risk perpetrators of violence. Similar patterns have been identified in other parts of Australia, including Sydney and Newcastle (Briscoe and Donnelly, 2001a & b).

By implication, workers in any business organization that deals with young male clients who are intoxicated will be at greater risk of violence, other crimes and/or 'swarming'. Hence, establishments where alcohol is served, that cater to younger customers (such as pin ball parlours), or where rival gang members may interact are at increased risk. However, *fears* of violent victimization have sparked a range of prevention strategies across the industrialized world.

Prevention of violence/bullying in the community

Escalating inequality in society has resulted in increasing spatial polarization of the population, including: (a) increasing segregation of the more affluent from the 'have nots' (such as in housing developments with high perimeter walls); (b) a growth in 'fortress' protection of many businesses; (c) avoidance of *perceived* risky times and places (many of the elderly avoid evening/night activities); and (d) 'demonization' of particular sub-groups of the population who are believed to be higher-risk.

The *fears* of victimization (rather than 'hard' data) appear to have spawned this emerging 'fortress' mentality. Those with the most to lose are increasingly segregating themselves into 'Crime prevention through environmental design' (CPTED) inspired gated housing communities (Martin, 2000) (see: CPTED discussion below). Conversely, where communities are sliding deeper into poverty, CPTED-based urban designs are deployed only on a minimalist *ad hoc* basis (Martin, 2000).

In public areas, software linked with closed circuit television (CCTV) can provide early warning of impending trouble. An evaluation trial of innovative software (called 'Intelligent pedestrian surveillance') began in two London underground stations in July 2003 (presumably fuelled by recent terrorist threats). If successful, this technology may overcome the 'overwork' required of security personnel who have to monitor multiple cameras and sites, and rapidly investigate behavioural anomalies, suspect packages, suicide attempts, and trespassers (see: Hogan, 2003).

The type of *lighting* outside shopping centres or in community areas can be selected to reduce site-specific risks. For example, increased lighting outside shops can minimize crimes against people and property (Sarkissian Associates and ACT government, 2000:12). However, enhanced street lighting may also attract young people at night time (Painter and Farrington, 1997: 225). To dissuade youngsters, a form of sodium lighting commonly called 'zit lighting' can be installed that accentuates pimples (White, 2000; White and Coventry, 2000). (Note: many lighting preventive measures have not been scientifically tested and therefore the results are not reliable.)

Local authorities and business owners may also dissuade teenagers loitering by changing the external environment so that it is an unattractive place to congregate (Sherman *et al*, 2002). The availability of seating provides one example. Because seating may attract groups of adolescents (as well as the elderly), the propensity for gangs to congregate may be reduced by removal of all seats in shopping precincts (Beck and Willis, 1995:96). *Music* has been used for similar purposes. Adolescents engaging in offensive behaviour may also be discouraged from loitering in a particular area that is not 'cool'. Strategies based on this principle include reports of classical music or Bing Crosby songs being played outside shops and on Sydney railway stations to deter young people (White, 2000; White and Coventry, 2000; Standing Committee on Law and Justice, 2000: 97).

Community safety groups can also play a seminal role in violence prevention strategies. Paul Kells from the Canadian *Safe Communities Foundation* argues that communities need to take ownership to reduce a range of safety threats, including violence (Kells, 2001). *Safe Community* programs involve a partnership between private and public sectors, work at both grassroots and policy levels, and target reduction of specific *injuries* prevalent in localized communities. Underlying the *Safe Communities'* innovation is the recognition that: (a) the cause of problems does not arise independently of the community within which an organization is situated; and (b) the most appropriate interventions can also be identified by that local population. Hence, local risk factors – which may vary across geographical areas – are the focus of attention. For example, road traffic deaths may be the prime target in one town, hold-ups in retail small businesses in another, or child cycle deaths in a different locality. Prevention strategies can become pervasive as individuals, small business owners, residents and community leaders collaboratively address local issues. Ananthaswamy (2003: 8) argued that casualties caused by guns could be considered a public health crisis and recommended tackling firearms access with the same vigour as infectious diseases.

Reduction of *precursor* risk factors appears to have had far less attention. For example, the building of social capital, housing-based neighbourhood stabilization, and the avoidance of segregated public housing may all be essential pre-requisites for the promotion of collective well-being/social cohesion and the reduction of resentment at relative deprivation (Kawachi *et al*, 1999).

When perceptions of relative deprivation are concentrated among populations in particular geographical areas, nearby businesses may be early targets. Hence some retail business owner/managers are likely to experience 'sentinel' events that predict future patterns of victimization in communities.

Occupational violence/bullying in retail small business

Because patterns of victimization vary considerably, and small businesses are widely dispersed, it is difficult to provide an overall 'picture' of the extent and types of crimes committed against them, including those involving violence. Small *retail* businesses can be victimized in a range of ways, including through hold-ups, break and enter, shoplifting, fraudulent demands from customers, credit card fraud, false invoicing, insolvent trading by clients who are later found to be unable to pay their bills, and perpetrators who initially masquerade as customers. Vulnerability is enhanced for sole traders, those operating in higher risk geographical areas (including remote regions), when there are few workers on site, and during evening/night trading. Technologically skilled perpetrators may skim personal details from stolen credit cards, commit illegal electronic cash transfers, or employees may execute inventory frauds. Those who operate a home-based business may be at increased risk of 'break and enter' accompanied by violence if cash, expensive computer equipment, or highly-sensitive data with potential resale value is on site.

It is important to recognize that violence is inter-related with other crimes against small business. For example: 'Shop theft is the main cause of violence and abuse...people who have an expensive alcohol or drug habit to feed or by gangs of youths after a thrill and some easy cash' (Hannett, 2002: 1). Thus, confronting a shoplifter may as frequently involve violence as do hold-ups.

Yet, small businesses are increasingly important for rural economies as business activity shifts from bigger to smaller towns, as seen in Britain over the past three decades. In a study involving 40 small businesses in rural and remote Britain, Marsh and Moohan (2003) reported widespread victimization, including theft of computers, vandalism and break and enter. While a number used sophisticated prevention strategies, at least one owner/manager relied upon the most basic of preventive strategies: '...stinging nettles and brambles around the boundaries...' for after-hours protection against break and enter (Marsh and Moohan, 2003: 46).

Two recent studies in Australia provide baseline information about levels of victimization experienced by small *retail* businesses. First, the Australian Institute of Criminology (AIC) and the Council of Small Business Organisations of Australia (COSBOA) conducted a nation-wide mailed survey. From the responding 4,315 small businesses it was estimated that around 50% were victimized over a 12-month period, with liquor outlets, service stations, general stores, pharmacies, and cafes/restaurants/takeaways most 'at risk' (Perrone, 2000; Taylor and Mayhew, 2002). Many small businesses suffered multiple incidents, particularly from shoplifting (89%), credit card fraud (62%), vandalism (60%), assault/threat/intimidation (57%), employee theft (52%), burglary (47%), and employee fraud (44%) (Perrone, 2000). Small businesses on arterial roads (71%), in residential zones (57%), within medical facilities (57%), and in isolated areas (50%) were reported to be most at risk (ibid).

A second study involving face-to-face interviews with fifty small *retail* business owner/managers confirmed that they were vulnerable to a range of crimes, including shoplifting, vandalism, 'break and enter', and armed hold-ups (Mayhew 2002b). However, there was a mismatch between *perceptions of risk* as compared with *actual* experiences. Notably, while small business owner/managers were very concerned about hold-ups, few experienced them. Conversely, while shoplifting and credit card fraud were very common, few owner/managers attached importance to the prevention of these offences (Mayhew, 2002b). In addition, intimidation and harassment from youths loitering outside retail shops were a major concern, with some valued customers too frightened to enter shops:

> A lot of [*named suburb*] people are old and frightened to walk down street; getting handbags grabbed and getting harassed in car park and around post office and asking for bus money. Pick on old ladies as they are frightened. They [offenders] are young.

Yet, loitering and exuberant behaviour by teenagers are not 'crimes' under the legal system in any democratic industrialized country. Nevertheless, provisions have been enacted to control *non-criminal* inappropriate behaviour in public places in some jurisdictions. For example, the Northern Territory Parliament of Australia enacted a Public Order and Anti-Social Conduct Act (2001: 5) which allows police greater powers if an officer 'has a reasonable apprehension that a person has engaged in, is engaging in or is about to engage in anti-social conduct' in or near a public place. Within this Act, 'anti-social conduct' included behaviour that 'interferes with trade or business at the place by unnecessarily obstructing, hindering or impeding a person entering at, or leaving the place ... disrupts peace or good order in the place or vicinity of the place' (ibid: 3). Thus, comprehensive prevention strategies may require commitment from business owner/managers, the local community, CPTED-inspired deterrence built into buildings, and a facilitating regulatory framework that can accommodate local risk factors.

Prevention of victimization of small retail businesses

The first step in any comprehensive prevention strategy is to identify effective means to reach the owner/managers of small businesses, which requires recognition of the distinct differences in their working lives and traditions compared with medium and larger sized organizations. Small business owner/managers tend to be entrepreneurial, work long hours, prefer concrete examples over abstract ideas, favour face-to-face contact over written communications (as they 'bin' much mailed information), and have a strong tradition of respecting advice from practically experienced people over civil servants or the academically qualified (Mayhew, 1997).

The prevention of a range of crimes against small business – including armed hold-ups – requires a basic understanding of insights from criminologists. For example, the best prevention against traditional business crimes (such as break and enter and armed hold-ups) may be through application of CPTED principles to site-specific risks. There are three core aspects to CPTED: 'target hardening', 'improved surveillance', and 'decreasing the rewards'.

'Target hardening' strategies can involve: time-delay locks on cash, 'drop' safes, improved locks, bullet resistant barriers or ascending ballistic screens, roof cavity protection, the installation of bollards outside entrances to prevent ram-raiding, and/or heavy shutters fitted to windows (Mayhew, 2000d). Technological devices may include: 24 hour perimeter and yard alarms, alarm activation points in key areas, cash dye bombs, entry/exit buzzers, and duress alarms fitted at the point-of-sale which may be silent internally but connected with security firms, other staff offices (or local police in a few countries). Changes can also be made to the internal layout, such as the point-of-sale counter raised so that it is too high for an average person to jump over, preferably with a raised floor height on the cash register side (Swanton and Webber, 1990: 22). These layout features increase the distance between workers (and the cash register) and potential offenders. Many small businesses also place emergency call numbers on all their phones.

'Improved surveillance' strategies are focused on increasing the probability that an offender will be identified, for example, by installing CCTV or digital camera recording. Shop displays can be arranged to allow a clear view from outside the store into the shop (signs in windows removed), and increased lighting at point-of-sale (Mayhew, 2000d; Swanton and Webber, 1990).

'Decreasing the rewards' strategies reduce incentives for robbery and armed hold-ups. For example, regular deposits should be made at banks and only a small 'float' retained in the till. Improved cash storage and transfer procedures are of core importance, particularly for small businesses operating in isolated areas, trading in the evening/night, and which have cash on hand.

A detailed small business check-list that assists with *self-auditing* of the risks of a range of crimes is attached to this book as Appendix 1. This *self-audit check list* has been field-tested with fifty small *retail* businesses during the course of an empirical research study (Mayhew, 2002b). This self-audit checklist should be of particular utility to those operating in rural and remote areas of Britain, Australia, Canada and the United States.

Risk reduction for the 'high-tech' crimes may be difficult for small business owner/managers who are not technologically skilled. 'Red flags' of potential criminal activity by employees of the business include a marked change in normal lifestyle, unusual spending patterns on credit cards, curious work-related expense claims, or atypical telephone call patterns (Grabosky and Duffield, 2001). At an organizational level, 'red flags' can include a lack of transparency in accounting systems, an absence of financial control systems, an escalation in credit card payments, or a sudden unexpected reversal of fortune (Grabosky and Duffield, 2001).

However, the costs of some prevention strategies may be prohibitive for economically marginal small businesses. In addition, as Sherman *et al* (2002) note, preventive strategies developed from experiences with traditional crimes may not be effective against new forms of crime such as internet-based scams. Nevertheless, some innovative technological developments are being standardized. For example, 'Crime Reduction through Product Design' strategies involve integration of protective technological features into products to: (a) reduce the ease of committing an offence, and (b) facilitate quick and effective responses following an offence. These security features are often combined with goods at the time of manufacture, for example, holographic images (or sub-surface laser marking or digital watermarks) imprinted on credit cards or polymer currency notes (Lester, 2001). Similarly, a program can be embedded in the hard drive of a laptop that identifies the computer identity and the telephone calling line at regular intervals to a monitoring centre whenever internet connections are made (Lester, 2001). The implementation of these technological innovations will assist small electronic and 'white goods' retailers, couriers, and those who rely on cash payments from customers.

However, there may be variations in the availability of preventive interventions between geographical areas. For example, small businesses in urban areas may have greater access to security services through sharing the cost of a security guard after-hours. Similarly, urban areas are more likely to have free community youth centres (which may provide an alternative to loitering around local businesses) than are available in small rural centres.

Security officers

Security officers perform a range of services for both business and community clients. Essentially their role is as a paid supplement for the police force. There are, however, three important differences: (a) the powers of security officers are quite restricted in terms of their capacity to restrain, evict or detain other people; (b) levels of training are less comprehensive; and (c) permission to carry firearms is usually far more restricted. As an aside, it is also important to note that police officers in some industrialized countries routinely carry firearms (e.g. Australia and US), but not in others (e.g. Britain).

The tasks security officers perform vary according to the client, and may include: patrol of business premises after-hours, safeguarding of retail shop strips, crowd control at sporting events, eviction of drunken or aggressive nightclub patrons, guard duties at banks, patrol on late-night trains, cash-in-transit escort, or at-hand support in hospitals at high-risk times (Somerville, 2001a). The people subjected to the attention of security officers may be: under the influence of alcohol or illicit substances, committing a crime, engaged in gang warfare, or have a mental health problem. Hence, security officers need training in a wide range of skills, including inter-personal techniques, anger de-escalation, legal rights and responsibilities, physical restraint, use of technological equipment, crowd control, emergency back-up procedures, and conflict resolution. In many jurisdictions, accreditation is now a pre-requisite to employment.

The role of *shared* locality-based security officers is of particular use to small businesses who cannot afford personalized high-cost protection from a range of crimes, including violent hold-ups. Thus, particularly in more isolated areas or at higher-risk functions or sites, the hiring of a security officer is a core aspect of an overall violence risk reduction strategy.

Conclusion

In this chapter, it was shown that the risk of experiencing a violent event in the community varied significantly between countries. Most at risk of a homicidal attack are citizens of South Africa, Russia and the Baltic states. Within Anglo countries, the citizens of the United States are at increased risk vis-à-vis those living in Australia, Britain, Canada, and New Zealand. There are a number of reasons for these variations in risk, including disparities in wealth between members of nation states, perceptions of relative deprivation, restriction of access to employment, desperation, intoxication, effectiveness of the criminal justice system, and differential access to weapons. For example, the available data appears to indicate that the comparatively easier access to firearms in North America and in more remote areas of Australia results in a proportionate increase in gun-related deaths.

The above discussions also indicated that the risk of occupational violence/bullying does not occur in isolation from the local community. While the *scientific* data are limited, all the indicators suggest that precursor risk factors for violence in a community can spill over onto workers and businesses within a locality. Individuals, community groups, business operators and workers are likely to be at increased risk in geographical areas where there are limited employment options, a concentration of higher-risk social groups, relative deprivation, few social supports, and increased levels of frustration among the population; all of which can be exacerbated by intoxication. Hence, the risk of occupational violence can vary significantly from one geographical area to another *within* as well as *between* nation states.

Public sector workers and retail small business operators may need to call on rapid support when these threats spill over into workplaces. However, there may be few back-up supports in more remote regions, such as the 'outback' of Australia, the north-west of Scotland, or the far north of Canada. Hence, those in remote areas may need to rely on local resources. Community organizations – such as the Canadian *Safe Communities Foundation* – enunciate a strategy which may be of enormous utility in reducing the incidence of such emergencies, in disseminating occupational violence/bullying prevention advice in higher-risk communities, and in assisting isolated small retail businesses in the adoption of preventive interventions.

In this chapter the risks faced by small *retail* businesses were examined as these organizations may well be the first to experience sentinel events that predict changing patterns of risk within particular localities. Reduction of the risks faced by small retail businesses – particularly armed hold-ups – requires interventions informed by criminologists, such as CPTED-inspired strategies. It is also important to identify the *obligations* of employers enshrined under statutory OSH legislation to prevent occupational violence (and, possibly, bullying). While there may be significant variations in regulatory requirements between nation states, there are commonalities (particularly between Britain and the Australia states where all *Acts* are based on the *Robens* blueprint). In particular, the *Duty of Care* obligations on owners/managers/duty holders require reduction of foreseeable risks through *prevention* in order to protect the OSH of employees and 'others' on-site.

To date, occupational violence/bullying prevention, OSH, and crime prevention have been seen to be separate spheres under separate jurisdictional departments, and prevention strategies have usually been developed in distinct ways. Nevertheless, the *safeguarding of organisations and their workers* requires promulgation of insights from all these disparate bodies of knowledge. Innovative links across disciplines, government departments, and industry and union bodies are likely to have significant positive outcomes.

Citizens of particular countries are probably entitled to succinct objective evidence so that they can make informed decisions (rather than emotive fear-based opinions), including:

- *Country-specific data* about homicide ratios and other violence indicators;
- *Localized* de-identified data;
- *Higher-risk* scenarios, and perpetrator demographic characteristics;
- *CPTED-based* interventions for small retail businesses (which can also be applied to home offices);
- *Effective intervention strategies* for individuals, matched against risk factors; and
- *Community-based* prevention strategies, such as Community Safety Groups.

If this information is collated, written in simple language, provided with practical examples, widely disseminated, and focused on *local* risk factors, the population as a whole is more likely to be able to objectively assess the risks. When local residents have an improved understanding of the risk factors in particular geographical areas/communities, they have an enhanced capacity to implement appropriate and effective prevention strategies.

In sum, the evidence indicates that the risk of homicide and physical assault varies between countries. Levels of violence also vary *within* as well as *between* nation states as the concentration of precursor risk factors alter across geographical areas. These risk factors include variations in cultural attitudes to aggression, norms of behaviour, social stressors, population demographics, ability to access weapons, levels of extreme frustration, efficacy of the criminal justice system, and extent of preventive interventions. Thus, tendencies to aggression spill over from local communities and present as occupational violence.

5

Occupational Violence, Psycho-Terror and Terrorism

Paul McCarthy

In the discussion that follows, the concept of psycho-terror provides a bridge that enables the investigation of common causes, and mutually furthering moments in workplace bullying/violence and global terrorism. The approach taken examines these phenomena through lenses of post traumatic stress, identity, organization, urbanization and gender, as well as lineages of victimization in Western thought.

Where occupational violence meets terrorism

The forces of global market capitalism that manifest psycho-terror in workplaces are also those that seed, incubate and provide conduits for terrorism. The global market forces that manifest workplace bullying/violence (McCarthy *et al*, 1995) also further the poverty, inequity and displacement that enables terrorism (Seabrook, 2002). Furthermore, workplace bullying/violence enables corruption, unfair trading practices, disrespect and disregard for diversity to continue (McCarthy *et al*, 2003). Risk factors for terrorism are likely to compound where global corporations enter partnerships with dominant national economic interests who sustain their positions through corruption and violence (see: Chua, 2003). One could reasonably conclude that corporations that fail to mainstream corporate social and environmental responsibility in management practices and do not implement policies and procedures to prevent workplace bullying/violence are likely to further the displacement, poverty and inequity in which terrorism takes root.

Those who experience intense threats as global market forces wash through their workplaces and communities and are unable to cope or to access justice often cling more tightly to work, cultural and religious traditions founded in violent reactions to oppressors. This tightening of grip on identity, status, role and beliefs exacerbates the sense of displacement and the feeling of trauma. In these circumstances, terror induced by perpetrators of workplace bullying/violence resonates with that from external threats in a dangerous world where perpetrators and victims cross over as cycles of violence ensue (Chomsky, 2001). A severe existential crisis manifests from compounding fears of economic, social, and cultural dying in an unforgiving universe. This terrifying experience motivates desire for re-affirmation that is acted out in clinging to traditions or the search for new meanings in work, organizational, social, political and cultural settings (Rylance, 2001).

Re-affirmation may be sought through many pathways, and access to these pathways and outcomes mediate the degree of violence in the victim's reactions. Where fear and hopelessness are overwhelming and remedies denied, resentment may be projected through traditional constructions of evil in targeting violence at perpetrators. Victims may also sacrifice themselves to the cause and suicide may enable relief from unbearable trauma. The suicide bomber conflates despair, revenge, politics and religiosity into a volatile mix for self-immolation in a war against evil.

The observation that one person's terrorist can be another's freedom fighter (Barker, 2003) is also true for the workplace in which the passionate advocate for employee's rights can be depicted as a disruptive force by others. Thus, in the politics of violence there is continuing debate about the 'reasonableness' of action that brings diverse ethical theories into play. In the microcosm of the workplace, debate about the 'reasonableness' of the actions of a perpetrator parallel macro debates about the ethics of the 'war on terror', detention, and curtailment of human rights. Unfortunately, utilitarian ethics of the greatest good for the greatest number can legitimize the psychic terrorization of those constructed as threat to the productive life of the group, the organization or the nation.

Dimensions of psycho-terror and terrorism

The concept 'psycho-terror' was first used to signify mobbing (or 'ganging up') by Leymann (1990) and was measured using the Leymann *Inventory of Psychological Terrorisation* (1997). The source of psycho-terror was located in the cumulative psychosomatic effects of repeated psychological abrasions experienced over months, sometimes years. Leymann (1990: 121) made particular mention of the possibility that the victim could react aggressively to the perpetrator/s and that such outbursts could be turned back against the victim in a stigmatizing process. The potential for fatal consequences through suicide or homicide was also noted.

The 11th September 2001 terrorist attacks on the World Trade Towers and the Pentagon via civilian aircraft have given new relevance to the notion of psychological terrorization. In the discussion that follows, terrorism is considered as an extreme 'external' form of occupational violence (CAL/OHSA, 1998). It is argued that the trauma of occupational violence/bullying is compounded by the psychosomatic impacts of incidents and intrusive memories of them, fears the events might be repeated, and subsequent difficulties in dealing with work and life issues. Researchers have found symptoms compatible with Post Traumatic Stress Disorder (PTSD) to be highly prevalent in victims of workplace bullying/ violence and terrorism (Leymann and Gustafsson, 1996; Mikkelsen and Einarsen, 2002; Einarsen and Mikkelsen, 2003; Schlenger *et al*, 2002).

PTSD triggered by terrorism radiates out from epicentre of an attack to persons within visual, auditory or olfactory range, as well as those tuned to communications that convey real time images and post-event commentary (Schlenger *et al*, 2002). Thus the traumatic effects of terrorist attacks spread widely from the meltdown of referents for security, safety and identity in target zones and beyond. Victims, witnesses and persons whose workmates, friends, family members or referents are injured by terrorist attacks may all experience traumas, directly and vicariously.

A national representative study conducted in the US by Schlenger *et al* (2002) (N = 3131) following the September 11 attacks found that 11.2% of respondents in the New York City Metropolitan area had 'probable' PTSD 1–2 months after the attacks. This finding compared with 2.7% in Washington DC, 3.6% in other major metropolitan areas and 4% in the rest of the country. The rate of reporting of symptoms consistent with PTSD was also positively correlated with extent of exposure to television viewing of the events. Considerably higher reporting of PTSD symptoms was evident amongst respondents with some relationship to victims and also to the military, although not at statistically significant levels in either case. That is, evidence indicates that geographical proximity and close links with victims was correlated with psychosomatic injury.

The destruction of psychological integrity

The ongoing experience of intrusive thoughts about incidents of occupational violence and the shattering of cognitive schemas and ostracization all interact in the destruction of the recipient's psychological integrity. The compounding effects of risk factors are outlined below.

Thought terror

The relationship between frequency of 'exposure to hostile workplace behaviours' and severity of impacts has been noted in a review of the work of several researchers completed by Keashley and Jagatic (2003: 53). They also observed that '... [e]ven seemingly minor behaviours can have significant negative effects when they occur frequently and over extended time periods' (2003: 53). The correlation between PTSD symptoms and exposure to bullying behaviours has also been confirmed by Einarsen *et al* (1999). Furthermore, the long duration of negative effects has been demonstrated by the findings of a study identifying symptoms consistent with PTSD in 65% of the victims even five years after the events (Einarsen *et al*, 1999). A heightened severity of impacts where the perpetrator's behaviour was experienced as personally degrading was also found in the study by Einarsen *et al* (1999). In addition, 73.6% of victims of bullying responding to a study by Mikkelsen and Einarsen (2002) reported that the quality of their interactions with family members and friends diminished, as did their leisure, household and sexual activities.

The reliving of the incident/s of mobbing through resurgent images, negative emotions, nightmares or reactions to stimuli has been found to be a key factor in the breakdown in psychological, physical, and social integrity of the victim. Such intrusive thoughts constituted the 'thought terror' that Leymann discerned at the centre of the PTSD reaction (Leymann and Gustafsson, 1996). Furthermore, symptoms of General Anxiety Disorder compatible with PTSD criteria compounded to impact more deeply in the deterioration of the victim's psychosomatic state over time in studies reported by Leymann and Gustafsson (1996).

Shattering of cognitive schemas

Basic cognitive schemas through which we derive our feelings of worth and regard from others and sense that the world is benevolent and meaningful are severely tested with the onset of PTSD following bullying/violence (Einarsen and Mikkelsen, 2003; Janoff-Bulmann, 1989, 1992; Lerner, 1970; Janoff-Bulmann and Freize, 1983; Epstein, 1985; and Janoff-Bulmann and Schwartzenberg, 1990). Exposure to aggression threatens the integrity and sense of invulnerability essential to effective functioning in the everyday life that is enabled by core belief schemas. The coming to awareness of the fragility of the basic assumptions underpinning identity and security in the encounter with bullying/violence can be painful for the recipient.

A severe crisis of 'cognitive disintegration' occurs and the victim struggles to revise and rebuild their core schemas so as to account for incongruent experiences. Those recipients who are unable to reformulate their world-view to maintain effective functioning are prone to experience '... a chronic state of cognitive confusion and anxiety that is characteristic of PTSD' (Einarsen and Mikkelsen, 2003: 137–138). A study finding that victims perceived themselves to be less lucky, capable or worthy and their world to be less caring, supportive, controllable or just, than did non-victims provided support for this hypothesis (Mikkelsen and Einarsen, 2002).

Ostracization

A 'socio-biological' explanation of the breakdown of psychological and physiological processes of persons who experience bullying/violence has been derived by Einarsen and Mikkelsen (2003: 139; with reference to Williams, 1997). Throughout history, the life or death dependency of the individual on the human group or tribe has meant that exclusion has the potential to evoke extreme anxieties. The description of the victim's experience of bullying as 'psychological drowning' (Einarsen *et al*, 1999) resonates with this explanation. Reasons for the destructive effects of the social ostracization identified by Einarsen and Mikkelsen (2003: 139–140) include the victim's experiences of:

- Extreme anxiety that interferes with the immune system and functioning of the brain with regard to aggression and depression (Williams, 1997);
- Threats to deep-seated needs for belonging, self esteem, capability and identity (Mikkelsen, 2001);
- Deprivation of the capacity to control outcomes of interactions or to respond to conflict, rendering targets powerless and suffering 'social death' (Williams, 1997);
- Pain, anxiety and worry as a consequence of failure to fulfil basic needs (see: Williams, 1997) that triggers '... extreme anxiety, depression, and in some cases even psychotic reactions', and '... desperate, erratic and sometime highly aggressive behaviours displayed by many victims of bullying' (Einarsen and Mikkelsen, 2003: 140; Einarsen *et al*, 1994);
- Diminished capacity to cope with demands of everyday living and working and display of aberrant behaviours by the victim that

prompt negative perceptions and actions by others (Leymann, 1990); and,
- Health crises precipitated by the decline in self-esteem and self-confidence, together with increasing anxiety.

Attribution, severity of impacts, and response

Attribution is a multi-factorial process in which the balance of blaming or assignment of responsibility by the recipient, and by others significant to them, can determine the extent of trauma experienced and shape responses. The web of possible attributions for psycho-terror extends outwards from the victim, to the perpetrator, the organization, and to wider socio-economic, cultural and political factors. The range of possible attributions on which a person can draw is determined by the explanatory systems they use in their social and cultural settings. A recipient's explanation of an experience of bullying/violence can influence the severity of impact. For example, a health professional that is slapped by a patient might experience a greater severity of impact if they attribute the patient's behaviour to maliciousness, rather than to dementia or some other locus of cause that diminishes the perpetrator's responsibility (Mayhew and Chappell, 2003).

Recipients can draw on a wide field of possible attributions in making sense of their experience of bullying/violence. The recipient's balance of attributions can also be influenced by those of their supporters, representatives and professional advisors, as well as by witnesses, managers, mediators and insurers. The provision of support to recipients of bullying/violence and the resolution of allegations often involve addressing misattributions. A possible field of attributions from which causal scenarios can be composed is set out in Table 5.1.

Incidents of bullying are often attributed to personality factors (Zapf and Gross, 2001). An early movement away from individualized attribution of causes of bullying/violence to location of responsibility with management as a work environment issue is evident in Leymann's (1990, 1996) work and in the Swedish Code *Victimization at Work* (1993). This move has continued to the present (Kennedy, 2001; Bowie, 2002; Ironside and Seifert, 2003; McCarthy *et al*, 1995, 2003a,b). Attribution of responsibility for occupational violence has tended to be in OSH and criminological terms for risk assessment and management. OSH has been particularly conscious of failure of managerial duty of care in regulatory terms.

Table 5.1 **Possible attributions for workplace bullying/violence**

Attribution	Basis of attribution
☐ Self-blame	Irresponsible or provocative behaviour by the recipient
☐ Accident	Unforeseen, unintended events
☐ Skills failure	Failure of interpersonal or management skills
☐ Personality	Predisposition due to genetic or social factors
☐ Role	Expression of role requirements in organizational culture
☐ Identity needs	To express command, power, or social status
☐ Abuse of power	Inappropriate exercise of positional power
☐ Manipulation	To achieve personal, group or organizational ends
☐ Greed	Unreasonable pursuit of personal profit or satisfaction
☐ Maliciousness	Intent to harm
☐ Revenge	Getting even
☐ Anger	Reaction to frustration or to unacceptable behaviour from others
☐ Sadism	Gaining pleasure from inflicting pain
☐ Criminality	Personal or social predisposition to criminal behaviour
☐ Mental health issue	Reaction to depression, PTSD, psychosis, alcohol or drug abuse, or dementia
☐ Disability	Intellectual or behavioural impairment, or learning difficulties
☐ Reasonable action	Reasonable management practice to achieve organizational goals
☐ System of work	Unreasonable work design, practices, resourcing, or training
☐ Work environment	Poor interactions, management, morale, commitment, or loyalty
☐ Normalised behaviours	The extent of bullying/violence that participants accept as necessary for normal system functioning or responses to external threats
☐ Social forces	Socialization to the use of bullying/violence, social fragmentation, or family breakdown
☐ Economic	Poverty, inequality, exploitation, or global market pressures
☐ Politics	Activism to influence decision making, change laws, achieve power or bring down the opposition
☐ Injustice	Breach of human rights, or civil or criminal regulations
☐ Cultural	Propensities to victimization and violence transmitted in cultural mores and clashes of cultural values
☐ Gender	Perpetrator and victim characteristics indicative of different risks for males and females
☐ Ethical	Breaches of codes of conduct, guidelines or moral norms
☐ Religious	The devil at work, eradication of evil, karma, ritual sacrificial violence, or clash of fundamental beliefs

Workplace bullying/violence has also been attributed to wider socio-economic, political and cultural forces, including systems and practices that threaten the livelihoods of the less powerful, brutal power plays, territorial wars, and appropriation of resources (McCarthy, 2003a). The attribution of bullying/violence to wider structural and cultural causes is evident in a study of Australian Aboriginal workers and managers by Williams and Thorpe (2003). The researchers documented the experiences of these workers and managers in dealing with racism, occupational illness and injuries, as well as emotional exhaustion in human services and obligatory community labour. Crimes of violence are often attributed to poverty, ignorance, oppression, abuse, and exploitation (Singer, 2002: 120). Class and caste formations have been found at the roots of systemic violence to those experiencing poverty as well as violent reactions (Seabrook, 2002). Innate drives to further the survival of the group or the gene pool have also been recognized as causes of aggression (Singer, 2002: 122).

Terrorist acts tend to be attributed to political, economic, cultural and religious conflicts (Ali, 2002; Hage, 2003; Chomsky, 2001; Barker, 2003; Jurgensmeyer, 2002). Anti-abortion terrorists who committed homicides in medical workplaces have attributed their behaviour to the Lord's will (Barker, 2003: 47). There is also evidence that the manner of attribution of terrorism mediates post-traumatic impacts (Schlenger *et al*, 2002; Norris *et al*, 2002). Those who are sympathetic to the terrorist's cause or see them as victims of conspiracies or negative economic and political forces may be less traumatized by incidents. The categorization of the perpetrator as 'evil' (Copjec, 1996) also brings moral and religious imperatives into play. A recipient who attributes bullying/violence to the perpetrator's departure from goodness inherent in the human condition may move on from the experience through forgiveness or reconciliation.

The discussion above indicates that bullying/violence can be attributed to many individual, organizational and wider social, economic, political and cultural causes. The next section focuses on the manner in which bullying/violence can be relayed through organizational roles in the interests of linking workplace bullying/violence and terrorism.

Transmission of brutality in organizational roles

The potential for 'normal' persons taking roles in organizational structures to transmit violence has been demonstrated in experiments by Milgram (1965) and Zimbardo (1988). In Milgram's experiment, members of the public placed in 'formal' organizational roles proceeded to follow instructions with considerable commitment. The instructions relayed by the participants produced apparent electric shocks with painful effects on recipients attached to electrodes. Unbeknown to the participants, the recipients were actors who simulated effects of the shocks. In Zimbardo's experiment, normal college students placed in roles of prison warders governing prisoners (also students acting in role) unleashed such violence that the experiment had to be cancelled. The ordinariness of those who relay real life terror in performing their organizational roles was evident in the family-loving normality of officials in the Nazi apparatus who were kind to their own families and to their next door Jewish neighbours (Bauman, 1991). Conroy (2000) also documented the banality of the lives of professional executioners and torturers. Experiences of 'executioners' of downsizing have also been detailed (Wright and Barling, 1998; Bowie, 2001).

Role positions that have the potential to relay bullying/violence are invariably at the centre of organizational operations as business is pursued, compliance with practices sought and external threats addressed. Arguably, in the flexible, multi-skilled work teams in the new organizations in the new economy, every employee is implicated in functions that can transmit systemic bullying/violence (see: Chapter 2). Role incumbents often implicate themselves in positions that relay violence because there are benefits to be derived. Threats of loss of role, for example through sacking, can also be a source of terror for those heavily invested in job security, consumption opportunities afforded by their work, and financial obligations (McCarthy, 2003a).

Identity terror

Processes in identity formation that inculcate predispositions to aggression and terror are examined in this section.

Violence in fault lines in identity

In our postmodern culture, identity and status are derived from exchanges around the display or vicarious experience of symbolic goods (Bordieu, 1884; Featherstone, 1988). Insight into the violence in these attachments is given by Lacan's (1977) understanding of the formation of identity in the mirror of the other, out of *lack*. The other is desired because they symbolize forms of enjoyment. Thus, self-identity is derived at the cost of suffering where the objects of desire are beyond reach or are prohibited. This 'pleasure in displeasure' was termed *jouissance* by Lacan, and pointed to painful tensions in the interplay of attraction and violence in identity formation (Lacan, 1977; Salecl, 1988). As Salecl (1998: 129) observes, '... [w]e hate others, because they have enjoyment', and this identity '... splits us from ourselves and from others'. Such sentiments are also evident in *Schadenfreude* in which the observation of violence to others affirms the well-being and good fortune of the observer (Portman, 2000).

Sadomasochistic moments

The conjoint inscription of pleasure and violence in identity-formation in postmodern culture and organization in global market capitalism also has its precursor in de Sade's (1968) 'matrix of maleficent molecules'. In de Sade's world, desire disposed itself '... towards such-and-such an object and against some other, depending on the amount of pleasure and pain I desire from these objects' (1968: 34). Arguably, the wholesale projection of desire into style in global markets in the information society unleashes this dynamic in economy, society and psyche (McCarthy, 1993). In these circumstances, identities that mirror symbolic goods ingest the violence bound up in their production. This violence is relayed through brutal management practices, lack of OSH standards, poor pay and conditions, environmental degradation, poverty, corruption, and child labour in third world and developing country workplaces (Seabrook, 2002).

This pleasure in pain inscribes fracture lines in identity that are severely tested where consumption opportunities are denied. Threats to consumption posed by bullying/violence and terrorism can open fault lines in identity and intense 'fears of falling' (Ehrenreich, 1990) that can form the basis for violent counter-reactions.

The politics of threat

Violence is also at the core of the delineation of the identity of the human group or organization. Insights into the psychological, political and religious moments that inscribe violence in organization have been given by Debray (1983). In Debray's terms, processes of delineation are at once: psychological (conferring identity and security); political (differentiating 'us' and 'them' or the enemy); and religious (overcoming fears of the future). Identity is wrought at the cost of cruelty, through initiation, norming, resource conflicts, and a veritable war between competing interests in a deregulating global marketplace. The tradition 'give me a good enemy and I will give you a good community' (Debray, 1983: 279) gains renewed relevance in the new organizational forms and identity politics that predominate post September 11.

Since the 11th September 2001 terrorist attacks, the politics of external threat has assumed renewed prominence beyond that which might have been expected with the decline of communism and the 'end of history' (Fukuyama, 1992). Huntington's (1996) 'clash of civilisations' has again been offered as an explanation for terrorism along Muslim and Judeo-Christian fault lines. Thus, the politics of threat resonates strongly through identity as an organizing force that conflates personal, economic, political, and religious motivations. The projective/introjective cycle of violence in identity formation outlined above provides fertile grounding for bullying/violence to arise in organizational life. It also provides conduits for bullying/violence to be projected into organizations from external sources and for organizations to perpetrate bullying/violence against others in their operating environments.

Gender sensitivities

Explanations for both workplace bullying/violence and global terrorism can be derived from patriarchy in the new global economy primed by the increasing entry of women to traditional male work and everyday life spaces. Women are implicated in violence as victims, witnesses, supporters, advocates for prevention, and sometimes as perpetrators. Prominence of women in the advocacy and development of measures to address bullying/violence in workplaces and schools has been notable, and has been located in women's experience of, and response to, a spectrum of violence that is predominantly male-initiated (McCarthy, 2001). Explanations for the predominance of women in discussing and reporting workplace bullying/violence and advocating for prevention can be made in the following terms:

- Concerns about bullying/violence at work have arisen in feminized cultures, i.e. cultures in which there are regulations to address employment rights that include women workers (Einarsen, 2000).
- Females entering part time and full time employment in growing numbers have brought with them an emotional literacy about bullying/violence. This literacy has been founded in the networking of resentments and advocacy of regulatory and policy responses to address domestic violence, sexual harassment, discrimination and stalking (McCarthy, 2001).

- The civil service has led the implementation of policies to protect women from violence in the community and in workplaces. Most public sector workplaces have sexual harassment, discrimination, gender equity and diversity policies. Notwithstanding these policies, women have been more commonly employed down the line and have found it difficult to break through the 'glass ceiling'. Thus, it is not surprising that reporting of incidences of bullying/violence has been higher in public sector than private enterprise workplaces; nor that women more often report experiencing bullying/mobbing than do males (e.g. Task Force on the Prevention of Workplace Bullying, 2001). Neither is it surprising that bullying is more often reported as coming down male dominated hierarchies (Zapf, 2001).

- Affirmative action policies have slowly increased the proportion of women in managerial positions, particularly in the civil service. However, gender stereotypes contribute to difficulties. Women, who exercise power aggressively, such as in ways that male managers characteristically do, risk being labelled as bullies or bitches or worse, by both male and female employees (Mapstone, 1998). Indeed, women may be more inclined to speak out about maltreatment from other women than do males about male perpetrators, due to female's socialization to express and work through emotions. In contrast, among males deference to other males up the hierarchy may reduce the risks of unleashing genetically programmed violence in a fight to the death. There is also the possibility that women might bully more proficiently than males in the new communication rich work teams due to their socialized verbal and emotional acuity.

- Gender stereotypes can also act perniciously to expose women to patriarchal violence. For example, due to their caring and nurturing stereotypes, women can be confined down the line in more domestic, supportive functions, away from line management where 'command' and 'toughness' are a prerequisite for advancement. Scutt (1994: 88) has also written that:

> ... the way women are stereotyped as all comforting, all succouring nurturers, or as depraved human beings driven wholly by their sexuality or 'womanness' was, it seemed to me, fundamental to the infliction on women of pain, injury and violence (Scutt, cited in Hockley, 1998: 107–8).

The history of syndromes of distress indicates that terror has long been visited upon women by male professionals (Showalter, 1997). For example, the diagnosis of hysteria legitimated an extraordinary range of psychotherapeutic and physical interventions. Little attention was given to the possibility that hysteria in women might be a manifestation of terror arising from powerlessness in male dominated society. Recent discourse about 'bullying at work' has also been placed in this lineage as a new signifier of distress in organizations impacted by globalization (McCarthy, 2003a). The discourse 'bullying at work' has given women a vocabulary to speak out about distressing experiences. However, remedies for the lack of power commonly associated with experiences of bullying/violence are rare. Rather than prevention, treatment is commonly prescribed in terms of psychotherapies, self-help, anti-depressants, and expensive legal actions for psychological injuries. Such remedies scarcely address prevention of embedded structural causes of the experience of workplace bullying/violence.

Terror associated with the entry of women into male-dominated workplaces has its precursor in the entry of women into male spaces in the city. Wilson (1991) identified female participation in mobs as a source of instability and revolutionary threat invoking paranoid fear in the nineteenth century city:

> There were women as well as men in the urban crowd. Indeed the crowd was increasingly invested with female characteristics, while retaining its association with criminals and minorities. The threatening masses were described in feminine terms: as hysterical, or in images of female instability and sexuality as a flood or a swamp. Like women, crowds were likely to rush to the extremes of emotion. As the rightwing theorist of the crowd, Le Bon put it, 'Crowds are like the sphinx of ancient fable; it is necessary to arrive at a solution of the problems offered by their psychology or to resign ourselves to being devoured by them.' At the heart of the urban labyrinth lurked not the Minotaur, a bull-like male monster, but the female Sphinx, the 'strangling' one, who was so called because she strangled all those who could not answer her riddle: female sexuality, womanhood out of control, lost nature, loss of identity (Wilson, 1991: 7).

Today, women and children are commonly identified as innocent victims of terrorism, and the kidnapping, rape and murder of women by terrorists has routinely fuelled the call to arms against terrorism. On the other hand, women have been supporters and lovers of terrorists, and have themselves been suicide bombers. Female interrogators have also been used to de-stabilize Al Qaeda suspects (Barker, 2003).

Speaking from her own experience as a member of the Weather Underground group in the US, Robin Morgan (2001, cited in Barker, 2003) locates the impetus to terrorism in '... the intersection of violence, eroticism and what is considered imasculinity'. To Morgan, the idea of the 'hero/martyr' who is violent, single-minded, controlling and self-sacrificing must be undone if action against terrorism is to succeed.

The potential for women to use their 'compassion and vision' against terrorism precipitated by patriarchy has been noted by Morgan (2001, cited in Barker 2003). However, women's stance against terrorism remains compromised by the weight of violence and exclusion to which they are subjected in patriarchal hierarchies. The extent of this violence is evident in an Amnesty International (2003) report finding violence against women – in homes, the community, work places, war and terrorism – as the most pervasive human rights challenge in the world today. Thus, there remains a need to stem this violence at all points to enable women's participation in those hierarchies through which patriarchal violence is transmitted and legitimated. For example, it is unlikely that the bullying/violence relayed down male dominated hierarchies and into decisions that fuel violence through degrading working conditions and communities can be addressed until women assume more prominence in governance in corporations, institutions, and public administration.

Urban terror

The experience of bullying/violence in the new organizational forms resonates with that in contemporary urban environs. The flexible work groups, networks, strategic alliances and labour practices that respond to deregulation and global market capitalism, wherein concerns about workplace bullying/violence arise, have their counterparts in the mixed use urban villages that mushroom in the new cityscapes and conurbations. Fears of violent crime (including assaults, rape, home invasions and a plethora of rages) have gained pace with divisions around wealth and poverty in the cityscape. These fears promote gated communities and neighbourhood-watch activities, and resonate with concerns about bullying/violence in the schoolyard and the workplace (McCarthy, 1998; 2003).

Insecurities experienced in workplaces and cityscapes undergoing restructuring are projected into the pursuit of wealth, rising property prices and security. In these circumstances, the investment in fear inscribes inequity and exclusion in public life (Caldeira, 1996; McCarthy, 2003). A struggle ensues wherein vested interests use violence:

> ... to make claims upon the city and use the city to make violent claims. They appropriate a space to which they then declare they belong; they violate a space which others claim (Holston and Appardurai, 1996: 202).

A similar struggle also ensues in organizations subjected to restructuring. Psycho-terror in the workplace resonates with that in the city and the quantum of fear drives the politics of mandatory detention of criminals, prostitutes, illegal immigrants, and terrorism suspects. The cumulative experience of threats, displacements and violations in restructuring at work and in the wider habitus contributes to psychoterror. Allegations of bullying/violence and advocacy for preventive measures are likewise used to make claims upon the organization. Where the sense of exclusion, inequity and violation are overwhelming and remedies are inaccessible, there are fertile grounds for political disaffection to mutate into terrorism.

Communal terror

The Hindu-Muslim conflict over the mosque at Ayodhya in India provides a fully developed illustration of the manner by which traditional mythos about violent histories is recycled into communal identity, economics, politics, and terror. Ayodhya remains a potent emblem of the rise of Hindu fundamentalism that swept the BJP into power in India in the last decade. The Hindu extremist's provocation was to demolish a mosque allegedly built on the site of a temple that marked the Hindu God Ram's birthplace. The following case study of 'Mala' illustrates capillary processes through which violence inscribed in traditional mythos is recycled in psycho-terror and terrorism.

Mala was an upper-caste New Delhi housewife who ran a small fashion export business from the garage in her house (McCarthy, 1994). Her daily routine involved purchasing materials in a nearby market place for delivery to her Muslim tailors. During one visit to the market a Hindu call to arms over the Ayodhya issue rang out a tape shop's loud speakers. Mala stopped, turned, listened intently, expressed interest in getting a copy, was visibly energized, and proceeded about her business with new focus and vigour. The words were those of a journalist:

> We Hindus have become a timid race, almost a cowardly race. We lack the courage of our convictions. Some of us don't even have convictions and have been trying to hide our shame under high sounding but empty phrases like 'secularism'. For the last so many centuries, the history of the Hindus has been created by non-Hindus, first the Moguls, then the British. Even today the Hindus are being denied the right to write their own history, which, to me, is like genocide. Until we write our own history, this land cannot be ours. The whole purpose of the Ayodhya movement is to change the history of India, nothing less, nothing more. For the first time in several centuries, the history of India is being made by Indians, call them Hindu, call them anything if the word sticks in your gullet as it did in Nehru's. Freedom does not mean flying your own flag or having your own government. *Freedom means making your own history, writing it in your own blood on the pages of Time* (Dubashi cited in Jack, 1990: 29).

The words in the tape seemed to penetrate Mala's soul and catalyse a vision that transcended all the difficulties that threatened to overwhelm her. Mala's religious and caste consciousness fed her chronic fear of violence. She had lost her father at an early age and that severely compromised the family future. Mala's husband had treated her unfairly and failed in his business ventures. She pursued the export business in her garage relentlessly since it offered the only hope for educating and caring for her children and ensuring the family future. While Mala killed off Muslims in her imagination, the call galvanized her purpose and destiny beyond all problems. Nonetheless, Mala's relations with her Muslim tailors remained remarkably cordial during this time. Mala's pathway away from fears of communal violence in India was in immigration to Sydney, Australia. There, Mala continues to react to fears of youth gangs of Middle Eastern origin, illegal immigrants, and terrorists that currently pervade discussion in the media and politics (e.g. Poynting, 2002).

However, blood flowed in the destruction of the mosque at Ayodhya in December 1992 with widespread looting, destruction, arson, raping and wounding of many, and killing of more than 2,000. In Bombay, the violence was also projected through fear, criminality, and anonymity in riots that changed the city-scape. Events there were depicted as '... nothing short of a revolutionary transformation – a geographical reorganisation of a complex city along communal lines' (Rhaman and Rattanani, 1993: 26). The areas in which this re-alignment occurred were 'the new communal tinder boxes' produced not just by differences but by 'the decay and criminalisation of the cities' (Chengappa, 1993a,b). As Chengappa and Menon observed:

> The concentration of political power, economic wealth and criminal gangs provides a highly volatile mix that is easy to ignite. Often as was seen in the post demolition riots, these became excuses to settle economic rivalries. In Calcutta, after December 6, property sharks instigated the destruction of a lower middle class Hindu colony in order to be able to construct a shopping complex on land later. In Gujarat, builders set fire to slums as a way of frightening off poor labourers and grabbing the Juggi land. In Kanpur, the superintendent of police found that the landlords were setting fire to their houses during the disorder so as to get rid of long-residing tenants who were paying only Rs. 10 per month rent (Chengappa and Menon, 1993: 28).

Anonymity in the city has also assisted criminal behaviours by young persons. For example, far greater violence was unleashed by the destruction of the Mosque at Ayodhya than in the villages (Shenaz cited in Chengappa and Menon, 1993: 28). In contrast to the villages, the lack of social roots, stable relationships, moral moorings, status and fear of stigma in the cities allowed participation in violence with fewer constraints.

Metcalf's (2002) tracing of the lineage of the Taliban back to Deoband formed in a village north of Delhi indicates a long trajectory of the conversion of the experience of victimization by Muslim's in local communities into terrorism. The Deoband sought to educate Muslim youth in the interests of community survival through alliances with more powerful Hindu interests. An offshoot of this Deoband movement engaged in the struggle over Pakistan's future in the final years of colonial rule and later positioned itself in the education of youth amongst the 3 million refugees that were located around the Pakistani border during the Soviet invasion of Afghanistan. These camps provided functionaries for the Taliban. Notably, these officers were provided with training and support by the CIA to push the Soviets out of Afghanistan. Thus the US was initially a resource for Al Qaeda.

Sacrificial violence

A long history of ritual sacrificial violence to assuage terrifying external forces marks the forging of human identity and organization (Girard, 1977). The terror of dismemberment and death through forces of nature and bestial predators motivated the mimetic sacrificial rituals of the Dionysiacs. When terror was projected out of immanence in nature into the removed Apollonian deities, sacrificial violence was used to enlist their favour. Mimesis of human sacrifice persists in Catholic ritual of communion involving the transubstantiation of the victim's blood into wine. It continues to transmit a powerful ethos of victimization and sacrificial atonement as the pathway to godliness. Along the way, this ethos has been operationalized to further secular global organizational ends in a lineage of sacrificial violence extending through the Crusades, the Inquisition, and colonialist genocide to the present day.

In the name of organization

Public torture, confessing, and execution of criminals have often been used in sacrificial affirmations of sovereign power throughout history (Foucault, 1977). The rise of the town, industry, the city-state, the nation together with parliamentary participation and civil society brought more sanitized sacrificial rituals. Resistance is now more commonly syphoned into conduits for the actualization of self, the community and the nation through the panopticon, bio-politics, finance, the corporation, the clinic, the school, the family, and other social and civil institutions (Foucault, 1991, 1981, 1977). Sacrificial violence is now democratized and privatized in productive life entailing endless micro-sacrifices, such as through overworking to accumulate symbolic goods and progress careers, as well as in gating of communities, enjoyment of debt, pursuit of fitness and self-help.

Rather than being tangential to contemporary workplaces, sacrificial violence can be identified at the centre of initiatives to bind employees to the organizational mission. Vicious scapegoating can be experienced by persons who take reasonable, lawful stances against organizational procedures or threaten professional solidarity and power (De Maria, 1999; Daniel, 1998). Public consultation processes are often conducted in ways that sense out and sacrifice the interests of those adversely affected by planning proposals who have less political influence (McCarthy, 1998).

Human sacrifice in Australia

Political and administrative expressions of sacrificial violence in the politics of external threat are evident in two key events during the 2001 Australian Federal election. In the 'children overboard' incident, refugees on a sinking boat held children aloft while begging to be rescued by an Australian patrol boat. Spin-doctors published pictures suggesting the immigrants threatened to throw their children overboard *if they were not rescued*. This lie was not revealed until a Senate Committee of Inquiry was established after the election. A Norwegian container vessel (the 'Tampa') observed international conventions in rescuing the boatload of refugees but was refused permission to offload them in Australian territory. The media sensationalizing of these incidents touched a deep-seated chord of fear and xenophobia that returned the conservative coalition to power. Similarly, in the lead in to the subsequent 2004 Australian Federal election, the politics of threat continues to sway a key cohort of globalization-disaffected voters (Megalogenis, 2002: 17; Manne, 2003: 19).

The politics of threat is also expressed in brutal administrative practices in the detention of refugees and of suspect terrorists that abrogate principles of justice in a social democracy (Mann, 2003; McCulloch, 2003). Such practices result in psycho-terror among detainees denied due process under international conventions and laws in social democracies. Psycho-terror is evident through the hopelessness, psychiatric trauma, self-harm and suicide attempts amongst such detainees in Australian detention centres and in Guantánamo Bay (Mann, 2003; International Committee of the Red Cross, 2003).

Sacrifice in the US

While so far not justified by verifiable weapons of mass destruction (WMD) threats, the post 11th September 2001 invasion of Iraq also addressed the possibility that Iraq might develop WMD that might be provided to terrorists. Benefits could also be derived from the development of the Iraqi economy and oil fields (Braude, 2003), removal of a threat to Israel, and the spread of democracy in the Middle East. Thus, rising casualties amongst American and other coalition forces and Iraqi civilians might seem worth the sacrifice. After all, some 30,000 gun deaths, and a greater number of road deaths per annum appeared 'normal' in US lifestyles. Human sacrifice to lifestyles and values can also be projected outwards, subtly, from the politics of oil. As Singer

(2002: 1) contends, the impact of emissions from gas-guzzling four wheel drives on climate changes will certainly induce more deaths from famine, rising seal levels and the spread of tropical diseases.

The 11th September 2001 attacks catalysed crises in the American dream into motivations for the invasion of Iraq according to Mailer (2003). The crisis trail was traced back into a 'powerful gene stream' predisposed to bullying/violence in workplaces and political life.

> The Republications who led the campaign to seize Florida in the year 2000 are descended from 125 years of lawyers and bankers with the cold nerve and fired-up greed to foreclose on many a widow's home or farm. Nor did these lawyers and bankers walk about suffused with guilt. They had the moral equivalent of Teflon in their soul. Church on Sunday, foreclose on Monday (Mailer, 2003: 3–4).

The rising tide of disquiet after the election was fed by several crises of confidence in addition to concerns about the political system identified by Mailer (2003). The stock market bubble had burst, and criminality was found at the heart of governance of the country's largest corporations and institutions (e.g. Enron, Tyco, the New York Stock Exchange). A Soviet mole was discovered in the FBI, paedophile priests in Catholic Church, and militant Christian's disgust at the moral depravity in the nation was unabated. Concerns about dependence on foreign oil and terrorism also contributed to vulnerabilities and fears. The ongoing use of 'evil' by George W. Bush was identified by Mailer (2003) as a 'narcotic pain-killer' for those most distressed, and war and the lure of World Empire beckoned as a way out of the conundrum. Consequent dangers of the assertion of fascism as 'a more natural state' over 'more perishable' democracy were of particular concern to Mailer (2003).

The resulting 'war on terror' was justified by the pursuit of freedom and security, yet there were warnings that it could be more destructive than the terrorism it was fighting (Chomsky, 2001; McCulloch, 2003). This inversion reminds us of Orwell's observation that fascism would come first to the West as freedom. The great majority of those who feel threatened by terrorism in the West agree to the sacrifice of freedom for security. The narrowing of the concerns of politics to security risks the prospect that '... security and terrorism may form a single deadly system in which they mutually justify and legitimate each other's actions' (Agamben, 2002).

Lineages of terror in the West

Eichmann's appeal to Kant's 'moral duty' in justifying his part in the genocide against the Jews in World War II (Bauman, 1991) reminds us to be wary about the perpetration of brutality in the name of freedom and the greater good. Such appeals have featured in justifications of state terrorism, cellular terrorism and the current 'war on terror'. The secularization of the Christian legacy in moral duty and in the construction of the enemy other, *the Jew*, can be found at the roots of Nazism and the Holocaust. The exercise of moral duty for the nation was also evident in rationalizations for the genocide of more than 150,000 in ethnic cleansing by Serbia in the 1990s, in which terror was perpetrated through 'killings, concentration camps and mass rape' that fell heavily on Muslims (Smith, 2003).

In reviewing the recent history of genocide, one could concur with Ash (1999) that '... [t]he Europe at the end of the twentieth century is quite as capable of barbarism as it was in the Holocaust of the mid-century'. The recurrence of such events in the West, and resurgent micro-fascisms prompts the following perverse question: What if the experience of terror was less a dysfunction, more a binding motivating force for the greater good, in the unfolding of Western traditions of Christianity and progress in modernity/post-modernity?

The Christian legacy

The celebration of the victim in Western culture has its roots in Christian mythos. In creation story, the fall of Adam and Eve stigmatizes every person as a victim who must claw their way back from hellfire or purgatory to redemption. Mutually furthering moments in the experience of terror, victimization, salvation, work and progress have been key factors in the ascendancy of Western reason, industry, politics and culture. This hegemony has been funded and disseminated through colonization and global market capitalism (Weber, 1930). *Orientalism* (Said, 1979) has been prominent in furthering this ascendancy and provided justifications for genocidal colonialism. Its lineage persists to the present day and surfaced in the genocide in Yugoslavia in the 1990s and the current war on terror.

In the microcosm of the organization, the Christian victim ethos can be discerned in the roots of PTSD as a living terror induced by bullying/violence. Psychotherapy, counselling, interpersonal skills, assertive training, anger management, and self-help now provide pathways to secular redemption for victims and perpetrators who have fallen from productive grace. Definitions of bullying/violence are more often inscribed in employment regulations. Checklists, guidelines and codes of conduct distributed by helping professionals, governmental agencies, and support groups provide identikits for victims and perpetrators and strategic options. Experts write medico-legal reports in terms of anxiety disorder, major depressive illness, PTSD and psychological injury to justify the victim's pursuit of compensation under workers' compensation, industrial, civil, and criminal laws. Treatment of victims with anti-depressants is routine (ABC, 2000). A bullying/violence industry has arisen around this service provision.

These constructions of the victim, perpetrators and therapeutic responses can be traced to the lineage of Christianity underpinning conceptions of justice, equity, and human rights in western social democracies. These values are also implicit in regulations governing entitlements at work in democratic nation states. By extension, the provision of definitions and remedies for workplace bullying/violence is an important part of the extension of humanitarian and civil values into the productive sphere. However, the question remains whether some victim identities/mentalities and remedies promulgated by the bullying/violence industry might work against the *collective* interests of recipients.

There are alternatives to the psychotherapeutic projection of impossible visions of 'love-full' communities (Rose, 1992) as a positive referent for the way things could be, beyond terror. Where access to remedies, typically requiring long time periods, require the victim display symptoms modelled in psychotherapeutic identikits and treatments, then the symptoms are more likely to become ingrained and self-furthering. For example, critical incident stress counselling may not always help traumatized persons move on from the experience. Instead, the counselling may expose the recipient to a vocabulary of psychosomatic effects and legitimation that enables psycho-terror to penetrate psychological defences. Nevertheless some recipients shrug off the impacts of workplace bullying/violence, or project their resentment into organized advocacy for prevention through a union, politics or activism.

Governmental self-regulative guidelines also provide a psychotherapeutic blueprint for the victim to confess the trauma induced by bullying/violence and for the perpetrator to acknowledge their inappropriate behaviour (Hatcher and McCarthy, 2002). The guidelines thus enact pastoral care in the interests of maintaining the health and safety of employees and the productive life of the nation (Foucault,

1981, 1991). The terror of the *fall* and redemption in Christian mythos is thus secularized in self-regulating organizations, self-managing work-teams, and self-helping individuals. In this schema, training employees in regulatory requirements and codes of conduct about workplace bullying/violence and in coping skills becomes obligatory.

From radical to modern evil

Investigation of the roots of modern evil led Copjec (1996) to identify formations in post-Enlightenment thought which seed both global terrorism and micro-fascism in contemporary workplaces and communities. This modern evil was depicted in the following terms.

> We are daily obliged to witness fresh atrocities as ethnic and racial hatreds seek to express themselves in the annihilation of their proponents' enemies; as nations decimate themselves, breaking into ever smaller and more fractious units; as terrorists of every stripe blow up people and buildings in an effort to protect their own and the rights of their 'unborn'; as multinational interests devise more advance forms of exploiting labour and crushing resistance; and on and on (Copjec, 1996: ix).

Such evil was difficult to constrain in a post-Enlightenment world since deconstruction and relativism obscured the authority for the law and morality. Copjec (1996) turned to the concept of 'radical evil' in Kant (1960) in diagnosing the roots of modern evil. Several thinkers informed Copjec's identification of contradictions in the relations of the subject, law and morality that provided windows for modern evil. If the basis of morality is located in human conscience, Freud indicated that the superego is not to be trusted as a source of guilt. Also, as Deleuze discerned, transgression was inevitably entailed in the exercise of the human will in striving for freedom and satisfaction. In these circumstances, moral law and the ethical subject manifested only as a negative point of reference for the transgressive exercise of the will. Kant considered that resistance to the actualization of moral law was responsible for the ineradicability of evil. Lacan also revealed our complicity in desiring what the other enjoys (Copjec, 1996: xvi, xi). The potential for others who stand in the way of our own enjoyment to be both sources and objects of terror is notable in this scenario.

In Copjec's (1996) reading of Kant, the propensity for radical evil is inscribed as the subject exercises its will, is alienated from itself and others, and transgresses referents for guilt in the pursuit of satisfaction, progress and freedom. Copjec considered that modern evil was a *response* to this internal fracturing of the will. Standards located within or without the will could become a source of servitude and terror. Where standards were inscribed in absolute bureaucratic terms, as in Kafka's totalitarian world, the experience of terror arises where citizens are forever charged with failures to follow obscure regulations, without clarification or right of reply (Copjec, 1996: xvii). On the other hand, our commitment to a self-referential system of surveillance placed the system of law and guilt in conduit, so realizing Foucault's panoptic self-surveillance in everyday life. Copjec pointed to Kant's warning against this 'voluntary yet intangible servitude', since it could lead to terrorization by the superego for the greater good (1996: xvii).

This 'servitude' brings Kant's conceptualization of radical evil face to face with modern evil in '... the colonialist taste for empire, the Nazi lust for genocide, and casuistry of the bloodiest kind' (Copjec, p. xvii). The exercise of evil in pursuit of absolute power, by torturers, executioners or perpetrators of genocide arises from the internal fracturing that alienates the will from itself. The only locus of resistance to this will to absolute power is in the victim, whether in life or dismemberment. Copjec (1996: xx) proposed that the apparatus that perpetrated modern evil (e.g. Nazism) was built not only on the projective construction of the threatening other, but also on *subreption* involving introjection of the 'supersensible' ideas of freedom and immortality.

Copjec (1996) found that this misrepresentation of freedom and immortality in Nazism, and in political theurgies since, lay in the unrealistic promise of infinite progress and pleasure. The problem with such a promise was its orientation to interminable cycles of becoming, failure and punishment, demanding 'greater sacrifices the more we sacrifice' (Copjec, 1996: xx–xxi). Copjec saw the ends of such sacrifice in Lacan:

> There is something profoundly masked in the critique of history we have experienced. This, re-enacting the most monstrous and supposedly superseded forms of the holocaust, is the drama of Nazism ... which only goes to show that the offering to obscure Gods of an object of sacrifice is something to which few subjects can resist succumbing, as if under some monstrous spell .... [T]he sacrifice signifies that, we try to find evidence for the presence of the desire of this Other that here I call *the dark God* (Lacan, 1977: 275).

The predisposition to assuage guilt in infinite progress was also explained by Freud's identification of the (figural) killing of the primal father in the ongoing introjection of murder, fear and obedience into human character while rendering the authority of the law inaccessible (Copjec, 1996: xxii). Copjec (1996) found a way beyond this tortuous treadmill in Kant's alignment of reason with action as good work justified by reference to a hypothetical universal final judgement of purpose and worth *made in the present*, not deferred.

Terror in modernism/postmodernism

Terror is unleashed in the unfolding of the modern/postmodern dynamic in the melting (Berman, 1982) of hitherto solid formations of meaning that had evolved over long periods of time as projective containers of fear of terrifying natural and social forces. A quantum revolution has ensued in which the clash of difference has been recognized as a key driver of meaning and organization. In the postmodern condition, language games have been played more or less violently, and terror has arisen where moves were made for pleasure, power or performativity (Lyotard, 1984). The fear of elimination from language games has been found at the roots of terror in the postmodern organizations and society.

The new nature in the postmodern condition has been termed *l'informatique* (Lyotard, 1984). It has been constituted by data banks, and particles of information that flow and swarm in an electronic force-field. In this condition, chaos theory has been found relevant in explaining field conditions for organizations (Peters, 1987). Flatter organizational forms, self-managing work teams, strategic alliances, networks, and tribal identity are among strategic responses proposed for managing ongoing threats to survival (Limerick and Cunnington, 1993; Clegg, 1990; Maffesoli, 1988).

Over-arching conceptualizations of the clash of meanings and interests in the evolution of meaning and organization have also been expressed by key post-structural theorists. For example, contradiction and resistance are central to Foucault's (1972) 'order of things'. The play of difference in language drives meaning in Derrida (1981). The 'molecular revolution' gives rise to new manifestations of violence in Guattari (1983). Particle flows of desire, intensities and plateaus are driven by attraction, swarming, repulsion and disintegration akin to quantum mechanics in Deleuze and Guattari (1987). Incommensurability in language games is central to a dynamic interplay of meaning and interest in Lyotard (1988). Also, the notion of 'force-field' (drawn from Adorno) governs gravitational pushes and pulls in Bernstein's (1991) 'new constellation'.

In the postmodern condition, violence in the clash of difference has been privatized and randomized in a manner that inexorably circumvents prevention and remains an ever-present motivating force. Potentials for confrontations and rage over resources, space and entitlements are compounded. Every person becomes a relay for violence in a plethora of rages and through group and political identities geared to threats from within and without. However, the deconstruction of ethical traditions is considered important in the progression to new ethics to account for violence in the clash of difference. For example, the roots of a positive postmodern ethics have been found in the terror of the deconstruction of unitary and monotheistic formations of meaning and the assertion of human potentiality in diversity (Rose, 1988).

In postmodernism, class struggle against systemic violence is replaced by competition between identity groups for access to consumption opportunities and respect for rights. The implication of the working class in these pursuits has led to their complicity with local and third world/developing country elites that perpetrate degrees of brutality in the production of symbolic goods. While those who experience workplace bullying/violence in western countries have access to guidelines and the law, those in poorer countries do not. These groundings of contemporary struggles for identity and respect leave class and caste differentials unexamined, and personal remedies are more often sought for '... humiliating effects of social and economic forces' (Seabrook, 2002: 115).

The speed-up in movement of information, people and goods through multiple interfaces in the global village increases the likelihood of abrasions in workplaces. Global information networks and the scale of movement of people and goods provide conduits for terrorism. In the US, Connolly (2003) points to daily movements across 7,500 miles of US borders and 160 entry points of: 50,000 trucks; 50,000 sealed containers (achieving a 2% inspection rate); 2,660 aircraft; 520 vessels; 348,000 vehicles and 1.3 million persons. In addition, there are some 39,000 commercial flights and 18,000 privately operated airports in the US. In addition, US government figures indicated that some 314,000 persons ordered to be deported had absconded, and that 93 million visas were issued in the 30 month period leading up to 11th September 2001 (Connolly, 2003). The policing of potential terrorists is difficult, perhaps impossible in this scale of movement. Global information networks also provide for service provision and an anti-terrorism industry has emerged, although its operation remains compromised due to the randomness of terrorist attacks.

The trajectory of postmodern terror is now played out in made-for-media destructive acts of biblical proportions directed at evil others (Jurgensmeyer, 2002). In the postmodern phantasmagoria we are all implicated in epic battles between good and evil. Indeed, a 'terrorist imagination' has been found in all of us, since we are prone to 'dream of the destruction of any power that has become hegemonic to that degree' (Baudrillard, 2001, cited in Barker, 2003: 16).

Mutation of resistance into global terrorist cells

The four phases in the mutation of resistance into terrorist cells depicted below have been composed less as a linear progression, more as mutually furthering moments in the mutation of resistance into terrorist cells.

Phase 1:   Historical incidents entailing violence and victimization engrave potent symbols around resistance to evil in cultural mythos. Refugees and immigrant labour in diasporas reproduce this mythos in the (re)imagining of ideal homelands and communities (Anderson, 1983).

Phase 2:   Blame for present day socio-economic, political and cultural crises is attributed to others construed in terms of traditional victim mythos (Jurgensmeyer, 2002). The victim's experience of the progressive degrading of their dignity, cultural beliefs, territorial rights and entitlements, participation in work and consumption, and self-determination produces aggressive counter-reactions. Lack of pathways to work through conflicts precipitates frustration and organized resistance.

Phase 3:   Global networking of resistance ensues through sympathetic nodes in the diaspora that assert identity through provision of moral and financial support for the struggle. Victim groups that act out resistance through violence may be further stigmatized as deviant and dangerous, isolated from participation in trade and political activities, and subjected to armed intervention (Chomsky, 2001; Said, 1979).

Phase 4:   Where justice is inaccessible, the experience of victimization unbearable, and open warfare not feasible, then resistance mutates into terrorist cells throughout cultural diasporas. Terrorist acts may be perpetrated at random locations with the aim of publicizing the terrorist's cause and their power, goading targets into violent responses, disrupting social, economic and political life, and breaking their opponent's will (Jurgensmeyer, 2002).

The form of the mutation of resistance into global cellular terrorism mirrors that of the flexible networked teams in the postmodern organizational forms that so destabilize local economies and cultural traditions. Notably, the Al Qaeda network was seeded and financed by wealthy educated immigrants in cultural diaspora who projected their situational resentment and nostalgia for homeland into alternative visions of modernity. The network also drew supporters from areas whose degradation was attributed to the evil other constructed in terms of mythos of subjugation and violence. Thus, cultural traditions and modernity have been fellow travellers in global terrorism. Family networks, inter-marriage, enlistment of the disadvantaged and the displaced in traditions of violent struggle through education, as well as modern skills, technologies, finance, transport, and communication systems were implicated in terrorism perpetrated by Al Qaeda.

To the extent global cellular terrorist networks mirror the organizational forms in the host they set out to destroy, they can be seen as a viral mutation that manifests in the progression of modernity/postmodernity into traditional cultures. Throughout history such mutations of resistance tend to be incorporated into the functioning of the host organization, community or nation over time. Victimization and violent protest have similarly marked the evolution of work, social, political and cultural practices and the transmutation of victim into perpetrator is common in the cycle. Historically, '... great persecutors are recruited amongst the martyrs of the not quiet beheaded' (Cioran in Bauman, 1993: 228). While the notion of victimization is premised in the ethical superiority of victims, it is also possible that victim's ethical claims may not be superior to those of their victimizers. It may be that the victims lack may lack the power that provides opportunities for them to perpetrate cruelty in the name of their own belief systems (Bauman, 1993: 228).

Conclusion

Propensities to experience and/or perpetrate terror are inscribed in the individual, the group, the organization, the culture and the nation through the pursuit of survival, identity, symbolic accumulation, performance, progress, freedom, the greater good, and immortality. The experience of terror and desires for violent revenge can arise from post traumatic effects of trauma, the reliving of incidents, fears incidents might be repeated, the shattering of cognitive schemas, and ostracization. Attributions of causes of bullying/violence can be made along a continuum spanning self-blame, behaviours of others, systemic factors, and wider socio-economic, political, cultural and religious forces. The manner of attribution of causes can mediate the severity of impacts of bullying/violence and the nature of responses.

Several fracture lines in identity formation that give rise to propensities to experience or perpetrate bullying/violence have been identified. These propensities are inscribed in identity in the splitting of the self from the other and the love/hate oscillation that ensues in seeking to enjoy as the other enjoys (Lacan, 1977; Salecl, 1998). Where identity is oriented to the pursuit of symbolic goods, threats to their consumption can precipitate experiences of bullying/violence and violent reactions against those who enjoy what we are denied. A deeper inscription of violence in identity has also been found in the pursuit of infinite progress, freedom, and immortality in our finite world in which groundings of the law and morality are inexorably deconstructed (Copjec, 1996). The impossibility of this quest, the misrepresentation of freedom and immortality within it, and the servitude of self to its ends in sacrificial violence for the greater good, have all been found at the roots of modern terror.

Globalization, or the unfolding of the modern/postmodern dynamics in global market capitalism, continually pressures the fault lines along which propensities to experience or perpetrate bullying/violence has been inscribed in identity. Terror is released in the relentless displacement of identity and projection of fears and threats through cultural traditions of subjugation and violent struggle. The politics of threat pervades, and with that the boundaries of identity, the group, the organizational, the community and the nation harden. This experience of threat seeds the perpetration of terror against the enemy within and without for the greater good. Global cellular terrorism can be seen as mutation of this cycle of resistance that can trigger a severe crisis in the hegemonic host.

The alignment of discourses of workplace bullying/violence and global terrorism indicates ways in which each of these forms of violence furthers the other. Global corporations are key conduits for the transmission of the systemic violence, identities, and economic, political, and cultural forces that degrade and antagonize local cultures to the point they become seedbeds for terrorism. Bullying/violence transmitted down and across male-dominated hierarchies is central to the mustering of corporate and local forces in productive and financial arrangements that antagonize local communities. From our positions of privilege in the west, workplace threats tend to be addressed through regulations that bind parties to codes of conduct and guidelines (Seabrook, 2002). This locus of prevention leaves global structural causes of displacement and impoverishment that ferment violence and terrorism untouched.

The experience of terror now justifies the exercise of bullying/violence for the greater good, for example in state terrorism and in the recent 'war on terror'. Apart from the exercise of overwhelming military force against sources of terror in foreign places, surveillance and detention have been pursued in ways that threaten liberties enshrined in social democratic constitutions. Those who oppose the 'dominant political orthodoxy' risk being tainted as fellow travellers with terrorists (McCulloch, 2003: 289).

The breaking of the cycle of workplace bullying/violence and terrorism depicted in the scenario outlined above would require a comprehensive web of international, national and local responses. An effective external locus of education, mediation of competing claims, and reporting of performance within an enforceable fabric of international OSH, human rights, environmental and criminal laws would be necessary to address this violence (Singer, 2002).

6
Occupational Violence/Bullying in the Health Industry

Claire Mayhew

This chapter is focused specifically on health workers' experiences of occupational violence/bullying. The discussion begins with an overview of the research evidence about the different forms of occupational violence experienced by various health professional groups, and then moves to an in-depth analysis of preventive strategies. The most appropriate starting point for the discussions in this chapter is with the only substantive international study of occupational violence in the health care industry ever conducted.

The international comparative study on violence in health

During 2001/2002 an international joint program on occupational violence in the health industry was established which involved representatives from the International Labour Office (ILO), International Council of Nurses (ICN), World Health Organisation (WHO), and Public Services International (PSI) (ILO/ICN/WHO/PSI, 2002a). The definition of occupational violence adopted by the ILO/ICN/WHO/PSI program was:

> Incidents where staff are abused, threatened or assaulted in circumstances related to their work, including commuting to and from work, involving an explicit or implicit challenge to their safety, well-being or health (Di Martino, 2002: 5).

Within this definition, 'psychological' violence was the term adopted by the joint program staff to encompass verbal abuse, bullying, sexual or racial harassment, mobbing, and threats.

The ILO/ICN/WHO/PSI committee members subsequently coordinated a series of research studies in six different countries focused on identifying baseline incidence ratios for a range of forms of occupational violence/bullying, including in Brazil, Bulgaria, Lebanon, Portugal, South Africa and Thailand. A concurrent Australian study was affiliated with the international project to enable comparative analysis (Mayhew and Chappell, 2003). Having used standardized methodologies and research instruments, these seven country case studies together provided an international baseline. As Di Martino stated:

> ... the country case studies provide, for the first time ever, a balanced, significant and reliable insight into the extent, patterns and impact of workplace violence within the health sector on a worldwide scale (Di Martino, 2002: 16).

The country case studies from the comparative international project on occupational violence in the health industry were subsequently published in a *Synthesis Report* by Di Martino (2002). This overview report identified relatively consistent patterns of occupational violence/bullying across countries, although there were some variations. Given the quite disparate cultures, levels of staffing, financial pressures, and preventive strategies adopted in the different countries, the similarities in risk factors and patterns are surprising. A brief overview of the findings is provided below:

- More than half of all health workers surveyed experienced at least one incident of physical or psychological violence in the previous year.
- The health occupational groups who experienced the highest levels of violence/bullying were ambulance officers, nurses and doctors.
- A significant correlation between exposure to violence and stress repercussions was noted.
- 'Psychological' violence was more prevalent than was physical violence.
- Patients were the predominant perpetrator group.
- Between 40% up to 70% of health workers who were exposed to at least one incident in the previous 12-month period reported significant symptoms of PTSD (Di Martino, 2002: 5–6)

There were also a few stark differences between health workers in the various countries:

- The estimated annual incidence ratio for *physical* violence events varied between countries: Portugal (3%), Lebanon (5.8%), Brazil

(6.4%), Bulgaria (7.5%), Thailand (10.5%), and South Africa (9% in the private sector and 17% in the public health service) (Di Martino, 2002: 25).

- The estimated annual incidence ratio for the different forms of '*psychological*' violence (as defined in the ILO/ICN/WHO/PSI project) also varied between countries. In the following subsections the data reflect health workers who experienced at least one event over the previous 12-month period:
- *Verbal abuse:* Portugal (27.4% in hospital, and 51% in health care complex), Bulgaria (32.2%), Brazil (39.5%), Lebanon (40.9%), Thailand (47.7%), South Africa (52% in private sector and 60.1% in public health service), and up to 67% in Australia.
- *Bullying and mobbing:* Australia (10.5%), Thailand (10.7%), Brazil (15.2%), Portugal (16.5% in hospital, and 23% in health care complex), South Africa (20.6%), Lebanon (22.1%), and Bulgaria (30.9%).
- *Racial harassment:* Thailand (0.7%), Bulgaria (2.2%), Lebanon (4.7%), Portugal (8% in hospital and 4% in health centre), and South Africa (22.5%).
- *Sexual harassment:* Bulgaria (0.8%), Thailand (1.9%), Lebanon (2.3%), Portugal (2.7% in hospital and 1% in health centre), and South Africa (4.6%) (Di Martino, 2002: 25–26).

That is, across all countries where the country case studies were conducted, health workers reported a significant level of occupational violence/bullying over the 12-month period of the study. The country-specific variations may have resulted from a significant spillover from the distinct societal and community pressures and local cultures (see: Chapter 4). These patterns are similar to the findings from British studies of general practitioners (see: Elston *et al*, 2002: 583). Workers in large hospitals in densely populated suburban areas with high crime figures were identified to be at particular risk (Di Martino, 2002: 29).

Several country surveys strongly hint at direct inroads of societal problems into the workplace with violence as one of the major issues. This exposure and permeability have always been known to be important factors in the triggering of workplace violence. It is now shown to be of extreme relevance as far as the health sector is concerned. The sector appears, by the very nature of the activities that are carried out therein, to be particularly exposed to external influences and as such, especially vulnerable to violence from the outside (Di Martino, 2002: 24).

In addition to the forms of occupational violence/bullying identified above, there is also an emerging new form of risk for health workers.

Prior to 11 September 2001, few health care workers would have recognized the risk of terrorist attacks on health workplaces, and ever fewer senior officers would have *planned to prevent* such incidents on home soil. The world has now changed. For example, on 30 December 2002, three US health workers (one medical officer and two administrative staff) at a missionary hospital in South Yemen were murdered during a terrorist attack. Subsequently, expatriate health care workers in a range of countries (including Afghanistan, Iraq, Israel, Pakistan, and Saudi Arabia) have been issued with precautionary warnings by their respective government authorities.

While at this stage the need for strict anti-terrorism precautions at all health care sites in all industrialized countries may not be necessary, wise administrators will at least be reviewing access arrangements, restricting vehicle movements close to buildings, and considering installing weapons screening devices at visitor/main entrances. Further, wise administrators recognize that health care workers are at the forefront of provision of treatment following terrorist bombings and bio-terrorism.

However, the primary purpose of this chapter is to focus on prevention of the more mundane (but still potentially fatal) violent incidents that appear to be increasing in both incidence and severity across the industrialized world, including in Britain, the United States, and Australian health care settings.

Variations in risk across health occupational groups

Overall, both the *incidence* and the *severity* of 'client-initiated' occupational violence (see: Chapter 2 typologies) in health care appear to be increasing over time. The fine-grain of the data gathered during the international country case studies indicated that there were variations in the incidence of the different types of occupational violence/bullying experienced across nation states, between hospitals and other health care centres, between the health occupational groups, and for diverse job tasks within specific professionals groups. In all countries involved in the international comparative study, 'client-initiated' occupational violence (as conceptualized by CAL/OSHA, 1998) was the predominant type.

Nevertheless, the evidence indicated that exposure to the risk of violence is not distributed homogenously across health occupational groups, or even within them. Rather, violence tends to be clustered around 'hot spots' (high-risk places) and 'hot tasks' (for example, when 'at risk' child patients are taken into government care), and is disproportionately perpetrated by 'hot people' (or those with higher-risk conditions e.g. drug and alcohol clients) (Mayhew and Chappell, 2003; Di Martino, 2002; Mayhew, 2000b; Brookes and Dunn, 1997). As a result of these diverse variations in risk, accurate knowledge of the risk factors, high-risk scenarios, and higher-risk perpetrator characteristics are essential for effective prevention at the local worksite level.

Community risk factors for violence

There are widespread indicators in the criminology literature that poverty, disadvantage, poor health and limited education are correlated with a range of violence indicators in the general community (see: Chapter 4; BCS&R, 2003a, b and c). These community indicators of distress may spillover onto health workers who provide services to injured and ill members of these social groups. Mullen (1997) has argued that a high unemployment level with marginalization of some groups provides the backdrop for violence directed at community support, health care, and other service industry workers. When cuts in public spending and services lead to a reduction in resources, clients may respond violently if they believe that they are being treated unjustly or are being discriminated against (Mullen 1997). Such scenarios may arise in the health care industry when public access to non-urgent services is restricted, there are very long waiting times, or financial contributions from patients/clients are increased (Chapter 4). Similarly, researchers in the USA have identified a clear link/spillover from community level violence into workplace aggression:

> ... workplace aggression is a partial outgrowth of community-level violence, and this has practical implications for the nature (that is, the level and content) of the interventions organizations use to prevent workplace aggression (Dietz *et al*, 2003: 323).

This community-level aggression is likely to be expressed through a range of indices, including domestic violence, on the street, in public places, and at local workplaces. There is also an increasing risk for health workers when offenders enter hospital premises for instrumental gain, or if opposing gang members are being treated simultaneously in emergency departments.

'External' occupational violence

If the risk of a spillover of community aggression into workplaces varies from region to region and suburb to suburb, it is likely that the risk of instrumental violence directed to health care worksites will also diverge. For example, in South Africa it was reported that:

> The public sector appears particularly vulnerable to violence with more crime-related incidents such as robberies, criminals hiding in big hospitals, gang wars being continued in the hospitals, patients with firearms and convicted criminals attacking the staff (Di Martino, 2002: 25).

As identified in Chapter 2, the criminological literature reports that there are four key risk factors associated with *'external'* occupational violence: exchange of money with customers, few workers on site, evening or night trading, and workers who have face-to-face communication with customers (Heskett, 1996: 16; Mayhew, 2002a). For example, during the course of an armed robbery perpetrators may engage in threats and physical assaults. In large health care complexes, café, post office and other retail outlets may be particularly vulnerable to violence that erupts during 'hold-ups'. Further, since a documented exponential rise in hold-ups of chemist shops in Britain and Australia over the past five years, health care sites/wards with drug cabinets must now be considered at significantly increased risk of *'external'* violence (Mayhew 2003b & c; Taylor and Mayhew, 2002).

> ... chemist shops and drug dealers represent the ultimate desperate robbery, for the ultimate object of the crime (drugs) can be taken directly ... offender described an armed robbery on a chemist shop where the offenders were 'shooting up' the stolen drugs as they drove away from the chemist shop ... the object of the robbery was to turn the proceeds into drugs as quickly as possible (Indermaur, 1995: 184).

> In [*operating*] theatre about 7 p.m. As I was coming to recovery room I saw guy with blood on face and looked like a drug addict and he had a jack crowbar and a long screwdriver. He was trying to break through the medicine cabinet but he released it when he saw me. I

asked him what was on e.g. plumbing problem. He ran out through sterilizing room ... I called security and they came after 20 minutes (Mayhew and Chappell, 2003: 31).

The risk of *'client-initiated' occupational* violence may also vary from one regional area and community to another.

'Client-initiated' occupational violence

As noted in chapter 2 of this book, the jobs at highest risk of 'client-initiated' violence in the US, Britain, and Australia are: police, security and prison guards, fire service, teachers, health care and social security workers (Mayhew, 2002a and 2000b; Fisher and Gunnison, 2001; Chappell and Di Martino, 2000). That is, workers who: '...provide care and services to people who are distressed, fearful, ill or incarcerated' (Warshaw and Messite, 1996: 999).

In the health care sector, the risks are not distributed equally between occupational groups. Broadly speaking, ambulance officers and nurses appear to have the highest incidence of victimization, and ancillary staff the least (with the important exceptions of security officers and admissions staff) (see: Di Martino, 2002; NHS, 2003 & 2000; Mayhew and Chappell, 2003). For example, Di Martino (2002: 28) reports that 'Ambulance staff would appear to pay the highest price'; and cites variable incidence ratios across countries that can reach 50% of the workforce for physical violence per year (in South Africa).

> Alcohol-induced outside hotels. You get a fair amount of abuse from public and/or family members. You get called to someone assaulted or injured; typically 95% at night and can be any day. Typically male but females do it as well; between 20 to 40 age group. We'll call for police to be on scene. Our job finishes after first aid or hospital transport. Personally happens once every 3 weeks. Depends on what level you call abuse. If someone tells me to go and get f..., or to piss off, I'd call that abuse – which happens every three weeks or more often (Mayhew and Chappell, 2003: unpublished interview transcript).

A slightly lower – but still unacceptably high – incidence for all forms of occupational violence was cited for nurses in each countries where the country-specific ILO/ICN/WHO/PSI case studies were conducted: Bulgaria (50%), South Africa (58%), Brazil (62%), Portugal (54% in hospital, and 74% in health centre), and Australia (67% verbally abused, 13% threatened, 10.5% bullied, 12% assaulted, and 5% some other form of violence) (Di Martino, 2002; Mayhew and Chappell, 2003).

Some *sites* are also at increased risk compared with others. The increased vulnerability of ambulance officers undoubtedly reflects the less protected and more unpredictable environments within which they conduct their job tasks, for example, treating the injured at motor vehicle crashes, or following pub brawls. Within hospital complexes, emergency departments and drug and alcohol clinics are higher risk sites for 'client-initiated' occupational violence (see: Brookes and Dunn, 1997). In addition, maternity and children's wards can be the site of high-tension events (Mayhew and Chappell, 2003). Further, British evidence cites an increased risk for those working with clients with mental health problems: the average number of incidents per 1,000 staff per month in 2001/2002 was 14 for all Trusts, and 33 for Mental Health trusts (NHS, 2003: 3).

The perpetrators of *'client-initiated'* occupational violence frequently share a number of characteristics. Clients who are intoxicated with licit or illicit substances, young males with a history of violence, and/or those who suffer psychosis or a neurological abnormality are higher-risk (Turnbull and Paterson, 1999). Australian research studies in the health care sector also indicate that perpetrators of 'client-initiated' violence are disproportionately male, younger, affected by substances or suffering from dementia (Mayhew and Chappell, 2003; Brookes and Dunn, 1997). For example:

'Guy in 30's; big bloke built like a tank who ripped the legs off a table in waiting room and legs of chairs – had nails in end. He swung them around and broke every television in the place. We kept him in locked up area; drunk. Police came (tied up at fight and took 15 minutes to arrive). All the patients left waiting room and we put security guards outside. He went to police custody ... (Mayhew and Chappell, 2003: unpublished interview transcript).

'Internal' occupational violence/bullying in the health industry

As noted in chapter 2 of this book, 'internal' violence/bullying (in the CAL/OSHA 1998 typology) is committed by individuals who have, or have previously had, an employment relationship with the organiza-tion. That is, the perpetrator and the recipient know each other, and may even have daily contact. The event may involve (a) a 'one-off' physical act of violence that results in a physical or emotional injury;

(b) initiation rites perpetrated on new employees, such as apprentices; or (c) some form of harassment or bullying that continues over time.

The international comparative case studies coordinated by the ILO/ICN/WHO/PSI indicated that significant numbers of health workers were at risk of bullying, racial and sexual harassment, mobbing, and verbal abuse from their colleagues (Di Martino, 2002). The highest levels reported were from South Africa, undoubtedly reflecting (to some extent) changing work practices and increasing ethnic diversity among the workforce as the remnants of apartheid are progressively dismantled over time. Within Australia, 10.5% of the sample of 400 health workers interviewed reported that they had been bullied via one method or another over the previous 12-month period (Mayhew and Chappell, 2003). Another Australian study involving 270 nurses in the state of Tasmania, reported that 30% were subjected to aggression on a daily or near daily basis from colleagues (Farrell, 1999). These estimates can be compared with the European all-industries estimate of 2% subjected to at least one *physical violence* event from fellow employees every year, and 10% subjected to bullying or harassment (Hoel *et al*, 2001: 21). Similar to the pattern of aggression from clients/patients, 'internal' violence/bullying directed to health worker colleagues is perpetrated along a continuum of severity, ranging from ridicule, to verbal abuse, to threats, to physical assaults (usually escalating in intensity over time) (Mayhew, 2000c).

The research evidence from a range of industry sectors indicates that bullying has an extensive impact on the health, well-being and productivity of immediate recipients, as well as on their co-workers (see: Chapter 8; Einarsen *et al*, 2003; McCarthy, 2001; Sheehan *et al*, 2001). Workers in the health care industry are likely to suffer similar negative outcomes to those experienced in other industry sectors. Their employers may also suffer significant legal and economic consequences, although these negative impacts are rarely recognized (Chapter 3). Given the current shortage of some groups of health care workers, any action to decrease the level of 'internal' violence/bullying is likely to have significant positive outcomes on recruitment and retention of valued staff.

The international research evidence also suggests that the *impact* of 'internal' violence may occur independently of the severity. That is, an on-going pattern of bullying can have a disproportionately severe impact vis-à-vis a 'one off' act of physical violence (see: second half of Chapter 8; Hoel *et al*, 2001; Einarsen *et al*, 1999). It is important to remember that the impact from 'internal' violence/bullying occurs in addition to the 'external' and 'client-initiated' forms of violence for health care workers inflicted by patients, their family members and visitors, and intruders.

Yes; almost daily verbal abuse. (a) <u>Verbal abuse</u>: daily from patients, typically drug-seeking or drug addictive, female, 20 to 30 i.e. young junkies. Usually directed to female nursing staff. (b) <u>Threats</u> from similar sorts of patients; males mostly; 20 to 30 to anyone who is there e.g. clerks, nurses, medical officers. (c) <u>Physical violence</u> – nothing major: punching, kicking, spitting. Not just those who are medically unwell. About once a month; usually male, some female, about 20 to 30 (Mayhew and Chappell, 2003: unpublished interview transcript).

Thus the cumulative burden may be significant and the consequences for individual recipients may be quite severe. The core challenge is to prevent the risks of violence arising in the first place.

Prevention of occupational violence/bullying in the health industry

All the indicators are that a *pre-planned, multi-faceted* and *organization-wide* approach has to be adopted to effectively prevent occupational violence. However, the available evidence suggests that the preventive strategies that are most likely to decrease the risk of occupational violence/bullying have *not* been widely adopted in the health industry. Arguably, one of the reasons for this inadequacy is that the occupational violence/bullying scientific research tends to be split into four quite distinct bodies of literature: (a) criminology; (b) occupational safety and health (OSH); (c) bullying; and (d) health. Very little overlap appears to have occurred. As a result, strategies that have worked very well in some industry sectors have not been taken up in the health industry (Mayhew, 2003c). For example, strategies that have been found to be effective in preventing 'external' violence associated with 'hold-ups' in the retail industry may be helpful in higher-risk health care sites such as methadone clinics (Mayhew, 2003b).

The management of risk (or 'risk control') is a central tenet of modern OSH regulatory regimes across the industrialized world, particularly those based on Robens-based regimes. As with other OSH hazards, the risk identification, assessment and control process has been found to be an effective strategy, as well as being a legal requirement in a number of countries.

Applying the OSH 'hierarchy of control' to occupational violence/bullying in health

Controls that eliminate the hazard altogether are the preferred option under these 'risk control' OSH regulatory regimes, for example, through designing out the risk when planning new premises or work systems. Where such architectural/engineering controls are not possible, or do not fully reduce the risks, administrative controls are recommended as additional strategies, for example, zero tolerance policies, rotation of staff to reduce the duration of exposure, or provision of information and training. In essence, under a 'risk control' protocol 'higher order' interventions such as elimination of the hazard are the preferred priority actions:

- Eliminate through design or engineering;
- Substitute;
- Enclosure;
- Administrative controls; and then
- Training

However, in general, it appears that occupational violence/bullying preventive interventions that are higher up the OSH 'hierarchy of control' have tended to be ignored in the health industry, and efforts appear to have been concentrated on 'lower-order' strategies such as improving the quality of training. A basic outline of 'higher order' OSH controls – as applied to the reduction of risk of occupational violence – appears below.

'Crime prevention through environmental design' (CPTED) applied to health

The OHS 'hierarchy of control' approach to managing hazards is consistent with the criminological construct of 'Crime prevention through environmental design' (see also: earlier discussion in Chapter 4). CPTED is effective in reducing the risk of both the 'external' and 'client-initiated' forms of occupational violence through *reduction of the opportunities* to commit violence or other crimes. CPTED is primarily accomplished through the work of architects, engineers, builders, and landscape gardeners who design-in risk-reduction features during the construction of a building or while refurbishments are in progress.

Risks are minimized by focusing on the placement of doors and windows, the immediate surroundings, and the selection of particular fittings and furniture (Caple, 2000; Mayhew, 2000d; NHS, 1997; Jeffery and Zahm, 1993; Clarke, 1992). For example, physical and symbolic barriers can be used to restrict vehicle movements, channel pedestrians along particular routes or corridors, as a form of access control, or to ensure all persons are screened for weapons before entering a building. These CPTED changes are usually long-term permanent features that do not need continuing financial support. Over time they may therefore be comparatively cheap if the cost of refurbishments is compared with economic losses from acts of violence/bullying that result in days lost, damaged property, and turnover of staff. CPTED strategies are usually tailored to site-specific risks for maximum benefit.

Specific strategies include 'target hardening' (making violence more difficult to execute), 'increased visibility' (to more easily identify perpetrators), and 'decrease the rewards' through better cash/drugs/valuables control that reduces the temptation of instrumental violence (Crowe and Adams, 1995).

Target hardening: Target hardening involves architectural or engineering designs (or re-designs) that control access to specific areas to make violence more difficult to commit e.g. deadlocks on areas where 'hot products' are stored (e.g. computers), reduced face-to-face contact during the supply of high-value goods (e.g. pharmacy), wider and higher counters at enquiry desks with raised floor height on staff side (e.g. in outpatients), designated safe escape rooms (e.g. in emergency departments), two exit doors in all interview rooms (e.g. in consultation/examination rooms), and key-card access to staff working areas.

The number of *entry points* to a building can be reduced, pathways can be designed to ensure pedestrian walkways are in clear view of people *routinely going about their business,* bollards can be placed outside pharmacy areas to prevent ram raids, bushes only planted outside where muggers are not likely to hide while they wait for victims, and narrow dark underpasses or lanes leading to public transport or hospital car parks designed out (Swanton and Webber, 1990).

Duress alarms: may be fitted at all desk areas and/or worn by all members of staff. Such alarms may be silent internally but link with computers that raise *automatic* emergency responses.

Increased visibility: Careful design of hospital/clinic buildings and their surroundings and fittings to increase visibility is a core component of an overall violence/bullying prevention strategy. The underlying belief is that if the risk of a perpetrator being caught is increased, this may act as a deterrent. Ideally, an activity that creates a risk of attracting a violent perpetrator is placed where many other potential witnesses routinely gather so that *surveillance* is almost automatic. For example, a methadone clinic might be placed in an area where numerous pedestrians pass by (This principle is routinely adopted when sites for ATM's are selected.) Similarly, interview rooms may be designed with large shatter-proof windows that are overlooked by numerous people routinely going about their normal health care tasks (Swanton and Webber, 1990).

The number of potential witnesses can be deliberately increased if high-risk times have been identified, such as Friday or Saturday nights in emergency departments (Mayhew, 2003a). The installation of closed-circuit television (CCTV) may also deter violence if potential perpetrators know they are being recorded. Prominent signs that video monitoring is on-going may help in this surveillance process.

Good lighting and high visibility are of increased importance during evening and night hours, including in hospital car parks, rarely used corridors and storage areas. The type and intensity of lighting can also be varied. For example, a form of sodium street lighting that accentuates pimples (colloquially known as zit lighting) has been successfully used to discourage young males from congregating in particular areas (see: Chapter 4; Painter and Farrington, 1997).

However, it is important to remember, that the 'increasing visibility' tactic is unlikely to work with those perpetrators of violence who are not rational, such as some dual diagnosis patients (The term 'dual diagnosis' is generally applied to clients who have been diagnosed with both a mental health and an alcohol/illicit drug diagnosis; as well as people with an intellectual disability and a substance abuse problem.)

Fittings and furniture: Wherever patients/clients have to wait for services, careful choice of furniture and fittings is crucial to reducing the risks. Waiting areas should be comfortable and spacious, and have a clear path to popular common-use fittings such as cold water dispensers and public access phones (CAL/OSHA, 1998). Ventilation and thermal controls are also important as violence may be more frequent at high temperatures (Chappell and Di Martino, 2000). Pastel colour schemes and soft furnishings can easily co-exist with CCTV, discreet alarm systems, chairs that are fixed to the floor (or too heavy to be lifted and thrown), and other CPTED features (Swanton and Webber, 1990).

Members of the public may have escalated anxiety levels if there are long waiting periods. Hence television and reading materials suitable for the particular social group (e.g. by age) can be provided. One important risk factor is the potential for misconceptions about queue jumpers when waiting patients/clients are not seen in the order when they arrive. Hence clear signs and explanations are needed for any delays in service provision e.g. in emergency departments. For this reason, new hospitals can be planned with a waiting room window over-looking an ambulance bay.

Counter design: The following design guidelines may be helpful, for example, in out-patient departments:

- Counters should be sufficiently wide that it is difficult for a patient/client to strike a worker (Swanton and Webber, 1990).
- Counters should be sufficiently high that it is difficult for an adult to climb or jump over (Mayhew, 2000b).
- The floor height can be raised on staff side so employees are higher than patients (Chappell ad Di Martino, 2000).
- A horizontal fixture can be built along the length of a counter about 165 cm from the floor (just below patient's head height) to reduce the risk of downward blows (CAL/OSHA, 1998).
- If there are a series of interviewing positions along a counter, vertical partitions between these may protect the privacy of disclosed information (Swanton and Webber, 1990).
- Duress alarms may be fitted at each counter (Swanton and Webber, 1990).
- Clear screens can be installed above counters with vertical slits to improve face-to-face communications and/or horizontal gaps to allow passage of documents. An additional counter with a lower desk/screen height is essential in hospital emergency and admission areas to ensure improved communication with people in wheelchairs.

Interview rooms: The following design features may be useful:

- Two doors in each room, with the staff member sitting close to one of these (CAL/OSHA, 1998).
- Patient/client access to interview rooms should be controlled (CAL/OSHA, 1998).
- A duress alarm needs to be discreetly fitted in each room (Mayhew, 2000b).
- Windows of shatterproof glass can be fitted so that clients and workers are in full view of all other people in the area (ibid).

- The furniture needs to be minimized, but sufficiently robust that it cannot be thrown or used to attack a worker (CAL/OSHA, 1998).
- Available equipment should be minimized e.g. staplers can be used as a weapon (Mayhew, 2000b).

A balance needs to be maintained between creating a relaxed environment and process of delivering the service to the client/patient, while ensuring the safety of staff *and other waiting clients*. This balance needs to be formally assessed through violence vulnerability audits, according to the degree and nature of the risks, the purpose of the site (e.g. emergency department vs. methadone clinic vs. aged care), and the level of additional risk emanating from the surrounding environment and activities (Mayhew, 2000b).

Health workers who are *off-site*

Those who work off-site or in community settings will need additional security. Global positioning systems (GPS) can be fitted to vehicles and linked with mobile phones, rapid key access to cars restricted to driver door only, and a series of administrative controls instituted such as detailed call-in systems with *inevitable* checks if no response is received (see: examples in Mayhew, 2000b). Prohibitions can be placed on off-site visits to previously violent clients, remote locations, and night/evening visits (ibid).

The best available evidence is that ambulance officers are at significant risk of occupational violence because they routinely work in more uncontrolled environments, as well in some isolated areas (see: Mayhew and Chappell, 2003; NHS, 2000). The existing risk management strategies in most health services should be regularly refined as the threats faced by ambulance officers increase over time, including innovative changes to vehicle design, emergency alarms, back-up systems, and protocols with police officers (which are already in place in most areas).

> We need a duress alarm on our person. At the present time, once we are away from our ambulance we have no way of singing out for help. They say we carry portable radios to call for help but they don't work 50% of the time. Plus if you are trying to protect yourself it's pretty hard to get your message across (Mayhew and Chappell, 2003: unpublished interview transcript).

In remote areas, the risks faced by health workers who work off-site are accentuated by limited resources, long distances between facilities/

towns, extended distances to drive, and potential delays in the arrival of back-up support personnel such as the police (see: Fisher *et al*, 1995; NH&MRC, 2002). Further, additional demands may be made by relatives of clients in more remote areas during crisis times (such as requests for accommodation) which when refused exacerbate levels of aggression (Mayhew and Chappell, 2003)

Reducing risks to health workers through administrative controls

Administrative controls are 'lower order' risk management techniques under the OSH 'hierarchy of control' prevention schema. Administrative controls focused on the reduction of occupational violence may include:

- CEO commitment to zero tolerance of violence that is *enforced* (see: NHS, 2003).
- Regular *violence vulnerability audits* of the site and work process (Long Island Coalition, 1996).
- Investigation and assessment of all reports and threats (Standing and Nicolini, 1997).
- Review of premises by crime prevention officers (Capozzoli and McVey, 1996).
- Correction of unsafe sites and processes (e.g. lighting).
- Security personnel in 'hot spots' (CAL/OSHA, 1998).
- *Rapid emergency response plan:* the adequacy of the rapid emergency response strategy to deal with threats to staff members needs to be regularly evaluated, as well as the extent to which this plan is understood, and rapidity of access to emergency contact numbers etc. (Mayhew, 2003c).
- Training and mock exercises for emergency response teams are needed, including for high-risk scenarios e.g. when rival gang members are being simultaneously treated in an emergency department (ibid).
- Ensuring that every purchasing officer in every health care setting is trained to have a comprehensive understanding of CPTED, and will refuse to 'sign-off' on the purchase of any buildings, refurbishments, or purchases (even of chairs) unless these are 'best practice' in terms of violence prevention (Mayhew, 2003a).
- Consideration needs to be given to allocation of the necessary powers for security officers to exclude or remove people from the site (Mayhew and Chappell, 2003).
- Regular maintenance testing of security devices (Mayhew and Chappell, 2003).

- 'Universal precautions' so that new staff and casuals are aware of risks and control strategies (CAL/OSHA, 1998).
- Multiple communication channels are needed to communicate with workers (Capozzoli and McVey, 1996). For example, violence prevention information may be attached to pay slips, or on notice boards in canteen areas. Additional efforts will be needed to reach casual staff.
- Regular training and re-training of all staff is needed to increase security awareness and enhance capacities. The content for a training program may include: risk factors for occupational violence/bullying, early warning signs, organizational policy and strategies, universal precautions, legal rights, progressive behaviour control strategies, and hands-on training with security hardware (Mayhew and Chappell, 2003; Mayhew, 2000b; CAL/OSHA, 1998). Paterson and Leadbetter (2002) have also provided detailed guidance on content and standards for occupational violence/bullying training programs. However, it is important to note that while the *severity* of incidents may be reduced by improved aggression de-escalation interpersonal skills, the *incidence* is unlikely to change.
- Development of relationships and protocols with local police.
- High quality data assists with tight targeting of prevention efforts. Hence good data are an essential pre-condition. Each database should include information on incidence in particular units, severity categories, perpetrator characteristics, possible causes/contributing factors, location features, other risk factors e.g. time of shift (Mayhew, 2003c).
- It is important that no penalties are attached to reporting (including undue time required to fill out forms). The establishment of a confidential electronic reporting system may be considered (Mayhew and Chappell, 2003).
- Consideration needs to be given to banning hospital/clinic *visitors* who are aggressive (NHS, 2003 & 2002).
- Flagging of aggressive patient files is increasingly being recognized as a core duty to workers, with progressive sanctions applied to serial perpetrators (see: NHS, 2002).
- For repeat offenders in smaller or more isolated towns, Apprehended Violence Orders (AVO's) can be taken out in the name of the health care facility/department rather than the individual victimized (as perpetrator and victim usually know each other's home as well as work addresses). The potential to access AVO's in an organizational name varies from country to country and state to state.

- There is evidence from both OSH and from criminology that the *certainty* of sanction on a perpetrator is a greater deterrent than is the *severity* of the sanction (Zdenkowski, 2000; Scholz and Gray, 1990).
- The human resource information system could record financial costs attributable to violence/bullying, similar to the ways by which workers' compensation costs can be separated out, e.g. flow-on expenses from absenteeism, sickness absence, workers' compensation insurance costs, turnover, short-term replacement of staff, diminished productivity, lost opportunity costs and poor reputation effects (see: Chapter 3).
- If the risks cannot be eliminated, then the organization of work can be reconfigured to reduce the risks, e.g. increased health worker numbers or security officers in 'hot spots' and at 'hot times' (Mayhew and Chappell, 2003; Chappell and Di Martino, 2000); and
- As in other industry sectors, independent evaluation of prevention strategies is essential, including worksite security.

To assist with prevention efforts at the international level, the ILO/ICN/WHO/PSI (2002b) released: *Framework Guidelines for Addressing Workplace Violence in the Health Sector.* These guidelines were developed to guide prevention initiatives at international, national and local levels while allowing adaptation to meet site-specific risk factors.

Conclusion

All the indicators are that the incidence and severity of occupational violence is increasing over time. The best available evidence is that *at least* 10% of health care workers will experience at least one incident of occupational violence/bullying each year. There are extensive direct and indirect costs to each health organization, to staff members, and to other client/patient groups as well as the local taxpayers. Hence wise health administrators will be prioritizing prevention strategies.

It has been argued in this chapter that a *pre-planned, multi-faceted and organization-wide* approach has to be adopted to safeguard workers and the health care organization from the risks of occupational violence/bullying. It is of core importance that the health industry begins to implement prevention strategies that are higher up the OSH 'hierarchy of control', including those known to be effective in other industry sectors such as CPTED. It is no longer sufficient to focus on violence training and emergency responses as these are not effective *prevention* strategies on their own.

In the health industry (as in other service industry sectors) it is important to note that obligations under the various OSH legislative frameworks in different countries are not generally diminished by the rights of patients/clients to their privacy. That is, CEO's have a duty of care to provide both a safe *place* and a safe *process* of work for their employees. In most jurisdictions, this non-delegable responsibility includes *everyone* on a site, including visitors and sub-contractors (see: Johnstone, 1999 and 1997). The prevention of *physical* occupational violence clearly falls under this obligation (as does, possibly, bullying).

Higher level controls must be implemented as a matter of urgency, not least because of the looming risks from terrorist attacks. A series of prevention strategies have been trialed and evaluated over a period of years in other industry sectors and have been found to be effective in reducing the risk of occupational violence/bullying. Yet these 'higher order' strategies are rarely understood and fully implemented across all health care sites in industrialized countries. Given requirements for prevention enshrined in OSH legislative frameworks, foreseeable risks of violence/bullying, increasing threats of terrorist attacks on home soil, and widespread knowledge about effective preventive strategies, there is no valid excuse to ignore the threats of occupational violence/bullying against health workers any longer.

Possible *safeguarding* strategies are elaborated further in Chapters 9 and 10 of this book including: the mutual intensification of risks for the different types of occupational violence/bullying, risk/severity spirals, and well as preventive interventions applicable across individual, organizational, societal and political levels.

7
Occupational Violence/Bullying in the Transport Industry

Claire Mayhew

Across the industrialized world the tasks completed by workers in the transport industry are highly diverse and as a result drivers face exposure to a wide range of risks. The working conditions of transport workers are also significantly different to those of most other workers. For example, in addition to the 'mobile' nature of their workplace, transport workers are employed in an industry where the reward system is frequently based on task completion rather than hours on-the-job, the 'command' structure tends to be a 'chain of responsibility' rather than a distinct employer/employee hierarchy, and the organization of work is generally less prescriptive than in most other occupations.

Transport workers may earn their living on land, at sea or in the air, share their workplace with unknown others (i.e. the road), travel at high or exasperatingly slow speeds, work at all hours of the day or night, regularly cross time zones, and/or not see or speak with colleagues for hours at a time. In addition, some transport workers are responsible for the lives of many passengers, and/or carry very high value cargo. A few deliver goods close to war zones. As a result of the highly variable nature of their job tasks, transport workers are exposed to diverse forms of occupational violence.

Forms of occupational violence/bullying experienced by transport workers

Some transport workers are at risk of 'external' violence in the course of hold-ups on-the-road (or piracy at sea), others to 'client-initiated' violence from passengers or customers (e.g. on buses, ferries or cruise liners), and many to 'internal' violence/bullying from co-workers or transport operators competing for lucrative contracts (see: typology in Chapter 2).

Unfortunately, little is known about the risks of violence for transport workers who have close face-to-face contact with customers, apart from taxi drivers (see: discussions below). There have been a number of anecdotal reports of bus drivers being 'held up' for small amounts of cash. For example, Somerville (2001b: 29) identified an increased level of risk associated with young alcohol-affected patrons late at night. There have also been a number of anecdotal reports of aggression from bus clients who are annoyed when schedules are not adhered to (for whatever reason), when fares evaders are apprehended, when tickets are provided through machines (which may not work well) and/or staff are not available at counters/service booths to answer questions.

Aggression from passengers in aircraft is an increasing risk. Williams (2000) has detailed some of the harassment, bullying and aggression that flight attendants are required to deal with in the course of their work; a risk that is enhanced by client intoxication during long-haul flights. In such scenarios, other passengers are also placed at risk. Since 11th September 2001, citizens of all industrialized countries are aware that such threats can be extreme if terrorists are on board airplanes.

More generally, little is known about the risks of violence for transport workers in public places, apart from those in close proximity to hotels and off-licences. Transport transit places (such as train and long distance coach stations) may be a prime target for armed robbery because of the retail outlets on site, extended hours of trading, limited security, and proximity to the haunts of drug dealers. An increased concentration of vulnerable commuters can also attract offenders to transport workplaces. For example, the popularity of jogging and cycling to work may place more people in situations where 'snatch and grab' robberies may occur; a similar risk is increasingly recognized by drivers of cars stopped at traffic lights. Similarly, congestion and gridlock on transport lanes may enhance proclivities to aggressive driving and road rage. Hence innovative urban transport system designs are needed that reduce vulnerability to a wide range of forms of occupational violence that can be experienced by diverse road users.

Nevertheless, the risks of occupational violence for transport workers cannot be divorced from the social cultures within which people live and work. Wolfgang and Feracutti (1969: 314–5) have argued that attitudes to violence are inculcated within subcultures of society so that in some scenarios violence becomes the normal and routine way of transacting interpersonal difficulties and conflicting wishes. A series of researchers have also identified close links between 'masculine' cultures and increased levels of violence. For example, Morgan (1997) has argued that aggressive driving behaviour (including road rage) can contribute to masculine identity through power games. Hence road rage can be located as one of many forms of aggressive and risky behaviour associated with status defense and enhancement. The core issue is, however, that other innocent drivers (including those driving taxis, courier vehicles, heavy haulage trucks, and buses carrying passengers) may suffer the consequences of roads being used as battlegrounds to test masculinity.

> For many in the 'at risk' group who are concerned with the presentation of their masculinity, driving becomes another arena of competition, struggle and apparent hierarchies of power. The road then becomes a particularly suitable 'screen' on which masculine power games are projected and played out (Morgan, 1997: 2).

There are a number of labour process variables that also influence the risk factors, including:

- All heavy vehicle, courier and taxi drivers share their worksite – the road – with non-workers and are at risk of road rage which is a form of violence that appears to be increasingly common (Mayhew and Quinlan, 2001).
- The majority of transport workers carrying freight are precariously employed (Quinlan, Mayhew and Bohle, 2001).
- Most transport workers work alone, and hence immediate support in times of crisis may not be available.
- Truck drivers and maritime workers transporting cargo are at risk of hold-ups, vehicle/vessel theft, or piracy at sea, particularly if they are carrying 'hot products' such as cigarettes, alcohol, lap-top computers, or other readily re-sold products that retain a high market value (Mayhew, 2001a).
- Drivers in the cash-in-transit sector carry large amounts of cash and are at particular risk of hold-up related violence – which has a comparatively high fatality rate in most industrialized countries (NOHSC, 1999).
- Those who have close contact with customers (such as bus/coach drivers), or who work with volatile persons (for example, on pub courtesy buses or even school buses), are also at risk of verbal abuse, threats and assault.

Thus the occupational violence experienced by transport workers can take many forms and be experienced on diverse sites. While a series of studies have been conducted in different sub-groups of the transport industry with the aim of identifying experiences of occupational violence, only limited anecdotal evidence has been gathered about transport worker offenders. Similarly, the interplay between violence expressed by passengers/clients/other road users and responding aggression from transport workers is poorly documented.

In this chapter, the occupational violence risks in the different transport industry sub-sectors will be identified, and the diverse risk factors distinguished. It is argued that the risks of occupational violence vary by occupational sub-group and according to transport task. The discussion below overviews the disparate risks for four core categories of transport workers: those involved with driving taxi vehicles, heavy haulage trucks, courier vehicles, and for workers in the maritime industry.

Taxi drivers

Violence is a common experience for taxi drivers, although most incidents involve only shouting, swearing and threats. Chappell and Di Martino (2000) estimate that the risks of violence for taxi drivers are around fifteen times that for all workers. Castillo and Jenkins (1994) also reported that taxi drivers had the highest rate of work-related homicide of any occupational group (26.9 per 100,000 workers) compared with the overall average of 0.71 per 100,000 workers. Taxi drivers face increased risks because they are often geographically isolated, have limited control over the passengers they accept, regularly work at high-risk times of the night and are also at risk of 'thrill kills' (Mayhew, 2000a). The risk of rape is an additional threat for female taxi drivers (see: Hume, 1995; Alexander *et al*, 1994).

The overall tendency across countries is for the *incidence* and the *severity* of occupational violence experienced by taxi drivers to increase over time. However, the risks vary somewhat between countries. In *Canada*, Stenning (1996) reported that 23% of taxi drivers experienced vandalism, 22% fare evasion, 15% 'other' theft, 10% robbery, 4% theft of taxi, 4% hijacking, and 3% some 'other' crimes. In the *US*, taxi drivers accounted for around nine per cent of all work-related homicides and had nearly sixty times the average rate of assault (NIOSH, 2000; NIOSH 1995; Jenkins, 1996). The risks were highest for unmarked 'gypsy' US cabs that go anywhere for a fare (Marosi, 1996). High levels of risk have also been identified in *Europe*. Elzinga (1996) estimated that around 75% of *Dutch* taxi drivers had been victimized in one way or another over a 12-month period, 22% had been threatened, and 4% suffered a physical injury. In *Australia*, a similar high level of risk has been identified in a series of studies in various geographical areas (see: Mayhew, 1999a; Haines, 1997; Keatsdale, 1995).

Importantly, even though these studies were commissioned by different bodies, conducted by a range of researchers in diverse countries, and assessed various risk factors, the findings were quite similar. However, in contrast to the data from the US, gun shots have been rarely reported in the European and Australian studies, with most incidents involving fists – and occasionally knives.

Forms of occupational violence experienced by taxi drivers

Taxi drivers are at significantly increased risk of 'external' violence (using the CAL/OSHA, 1998 typology described in Chapter 2). The reasons for the increased vulnerability of taxi drivers are patently clear: they carry cash and are a relatively easy target. Virtually any potential offender can 'hail' a taxi from the street and identify a destination for 'drop off' where there are few potential witnesses. Media publicity may, unfortunately, increase the possibility of future attacks as these may mention – or even highlight – the fact that the victimized taxi driver was robbed of cash secreted within the cab or in his/her wallet.

In almost all scenarios, taxi drivers have close contact with their client/customer. Hence, taxi drivers are at high risk of 'client-initiated' violence in the form of verbal abuse, threats and even assault.

'Internal' violence/bullying may arise when few passengers are requiring transportation and drivers are competing for fares. Nonetheless, the available evidence indicates that this form of occupational violence is relatively rare among taxi drivers.

Empirical evidence on experiences of occupational violence among taxi drivers

The findings from Australian studies into occupational violence among taxi drivers are consistent across studies. Substantive studies have been conducted in three Australian states over the past decade. The types of injury are consistent with attacks from passengers in rear seats, with the majority experiencing upper-body lacerations/bruising/fractures. In the study conducted in the Australian state of Queensland, 81% were abused in the immediately previous 12-month period, 17% threatened, and 10% assaulted (Mayhew, 1999a). The study conducted in the state of Victoria reported that 84% of taxi drivers were victimized through fare evasion, 79% were abused, 40% assaulted, and 24% robbed (Haines, 1997). In the study conducted in the state of New South Wales, 69% of taxi drivers experienced fare evasion, 44% were abused, 19% threatened, 15% assaulted, 11% robbed, and 4% victimized via robbery with violence (Keatsdale, 1995). The variations between these studies may reflect a wide range of differences in offender characteristics, population variations (see: Chapter 4), size of cities, levels of poverty/stress in the local community, preventive measures in place, and/or even levels of drug addiction in the general population. Nevertheless, in each of the three Australian states where quite separate studies were conducted, *at least* 44% of taxi drivers were abused and 10% assaulted over a 12-month period.

Many incidents were perpetrated by aggressive and inebriated young males – characteristics that are consistently reported in other criminological studies (see: Chapter 4; Chappell and Di Martino, 2000; Wolfgang and Feracutti, 1969). The core risk factors for taxi drivers have been identified:

- Male passengers;
- Younger clients/customers;
- Working in the evening or night time;
- Intoxication of passengers with alcohol or illicit substances;
- A 'hail' from the street (rather than a phone booking for the taxi);
- Inner-city pick-up point;
- Disadvantaged socio-economic clients;
- Lower socio-economic pick-up/drop-off suburbs; and
- The pursuit of fare evaders ('runners') by the taxi driver (Mayhew, 2000a and 1999a).

Table 7.1   **Patterns of occupational violence reported in Australian taxi driver studies**

	Queensland	NSW	Victoria
Form of violence in past 12 months			
Abused	81%	44.4%	79%
Threatened	17%	19.3%	–
Assaulted	10%	15.2%	40%
Robbery involving violence	–	3.8%	–
Robbery	–	11%	24%
Fare evasion	–	69.4%	84%

(Mayhew, 2000a: 3; Keatsdale, 1995: 27; Haines, 1997: 9–15).

Passengers presenting with a greater number of the risk factors appeared to be more likely to commit occupational violence.

Discussion and summation

The overall conclusion from the available scientific studies seems inescapable: low-level occupational violence is endemic in the taxi industry across the industrialized world. Verbal abuse may even be a daily occurrence, with threats and assaults also comparatively common vis-à-vis other occupational groups. Few of the incidents cited in the Mayhew (2000a), Haines (1997) or Keatsdale (1995) studies were formally reported to the authorities in the respective Australian states, sometimes because the self-employed had no mechanism to report, because the resulting injuries were not *physically* severe (although emotionally debilitating), and/or because of job loss fears. Employment status has previously been shown to be an important mediating variable that affects capacity to report (particularly among the self-employed), and complicated bureaucratic procedures that may reap few financial rewards are a further disincentive to formal reporting of less severe violent incidents (Quinlan and Mayhew, 1999). Further, across all countries where substantive studies have been conducted, the data indicate an unequivocal increase in the level of risk over time.

Prevention strategies that directly address the identified risk factors have been progressively implemented in taxi vehicles across the industrialized world, although the costs of their introduction are generally hotly contested. These interventions may include the fitting of safety protective screens between driver and passengers, emergency lights, duress alarms, security cameras, passenger restraints, driver controlled door locks, cash reduction strategies, and inter-personal skills training (see: Mayhew, 2000e). Further, the regulatory frameworks in diverse OSH jurisdictions/authorities have increasingly been modified to explicitly identify the interior of a taxi as a workplace. This amendment clearly brings taxi drivers under the protection of the 'duty of care' provisions enshrined in all Robens-based OSH legislative frameworks (Easton, 1997: 157).

In sum, taxi drivers are 'popular' targets for 'external' and 'client-initiated' violence because they work alone, are relatively unprotected, accept passengers whose attitudes to violence are usually unknown, and they almost inevitably carry a supply of cash from previous passengers or to provide change for larger notes proffered by paying clients. The risk factors for occupational violence/bullying in heavy haulage transport are quite different.

Heavy haulage transport

The risk of occupational violence for heavy vehicle transport drivers is not widely recognized, and few scientific research studies have been conducted in industrialized countries. However, valid indicators of the risk factors can be gleaned from a range of research studies, many of which were commissioned for other purposes.

Forms of occupational violence experienced by truck drivers

Heavy vehicle transport workers face the risk of 'external' violence during 'hold-ups' for cargo or theft of a truck. Worldwide, cargo losses have been estimated at $US30 billion a year with road transport accounting for around 87% of the direct-cost value of lost cargo (Salkin, 1999; DeGeneste and Sullivan, 1994). In contrast to many other business crimes, cargo theft often occurs during business hours, with large-scale theft at freight-forwarding yards or on highways frequently involving collusion between 'insiders' who alert potential offenders of opportunities. The typical gang involves: (a) an inside informer who supplies cargo and schedule information, (b) a 'spotter' who follows the truck and operates as the 'lookout', (c) an offender who holds the vehicle up and who may drive the truck away to another site for cargo stripping or vehicle 're-birthing', and (d) a 'fence' who quickly disposes of the stolen goods (see: Mayhew, 2001a; Salzano and Hartman, 1997; Small, 1998). In addition, heavy haulage drivers (particularly those who transport cargo internationally or carry containers from overseas ports) may inadvertently become the targets of drug or arms traffickers who secrete illicit goods within containers or packages. When such illicit cargo is retrieved, the truck driver may be at the mercy of violent offenders who 'hold-up' the vehicle for the stashed illicit substances or goods.

'Client-initiated' violence (as defined in the CAL/OSHA typology – see Chapter 2) is experienced by heavy vehicle drivers when customers are annoyed at delivery charges or delayed time schedules. However, there is very limited evidence of these incidents, and hence it must be presumed that this form of occupational violence is usually of lower severity and/or a comparatively rare experience for heavy haulage drivers.

'Internal' violence may arise when owner/drivers under economic stress under-cut prices in a very competitive industry. For example, tensions may rise between those jockeying for preferential loading/ unloading places in long queues at freight forwarding depots. Such scenarios may fuel aggression if there are long waiting times which are not part of paid working hours. In environments where excessive hours of work and/or shiftwork are common, the potential for stress-related aggression almost inevitably rises.

In addition to the three 'common' forms of occupational violence experienced by many workers, other road users can engage in aggressive and dangerous conduct that threatens the safety of heavy vehicle drivers. The road rage experiences of heavy haulage drivers can include the throwing of missiles towards trucks, tailgating, verbal abuse and/or rude hand gestures. Even such lower-level forms of occupational violence can be distressing. For example, one heavy vehicle driver who transported battery hens to a chicken processing plant was routinely abused by animal rights sympathizers: '... young women will abuse us for being cruel as we've got live chickens on board ...' (Mayhew and Quinlan, 2000: interviewee 261). It is important to note that the negative emotional and physical consequences following all forms of occupational violence occur in addition to extensive other stressors experienced by the majority of transport workers (see: de Croon *et al*, 2002).

Empirical evidence on experiences of occupational violence among truck drivers

Three research studies have been conducted by the author of this chapter that together provide indicative data on the extent and severity of occupational violence experienced by heavy vehicle haulage drivers in Australia. First, a study was conducted that involved interviews with 71 drivers; these data were then compared against official workers' compensation claims, hospital treatments and fatal road transport crash data (Mayhew, 1999b). Second, the author coordinated a study that compared the OSH experiences of employees in the transport industry against those who were contractors/subcontractors (Mayhew *et al*, 1996). Third, in 2000 the author of this chapter conducted 300 face-to-face interviews with long-haul truck drivers as part of a formal government *Inquiry* into this industry sector (Mayhew and Quinlan, 2000). Since there was no conceivable way by which to obtain a randomized sample of all long-haul drivers, this sample was selected in as systematic a manner as was possible: interviews were conducted at freight forwarding yards, 'truck stops' along major interstate highways (which sell fast-food and fuel to truck drivers), and in truck parking areas.

In the final study, fully 33% of heavy transport drivers interviewed were verbally abused in the immediately previous 12-month period,

21% experienced road violence (in the form of road rage), 8% were threatened, and 1% was physically assaulted (Mayhew and Quinlan, 2001: 39). In the absence of other substantive studies, these data can be taken as baseline incidence ratios. The road violence had the most severe potential consequences (e.g. when missiles such as bricks were thrown at windscreens) and most commonly occurred in heavy traffic situations near red lights or roundabouts, or when heavily laden vehicles progressed slowly up hills.

'Client-initiated' violence from customers followed delays in delivery, or unexpected price/charge increases. For clients whose own business was only marginally viable, these additional strains could threaten their survival and compound already stressful working situations.

Incidents of 'internal' violence/bullying appeared to be fuelled by very tight economic pressures, delays in loading/unloading vehicles, and overworked and fatigued drivers. In freight forwarding yards, loading delays exacerbated the tensions as waiting time was usually unpaid time for self-employed and owner/operators of heavy haulage vehicles. Hence delays, queue-jumping, mistakes by forklift drivers, time pressures, and tight delivery schedules all heightened tensions at peak loading/unloading times and fuelled aggression (Mayhew and Quinlan, 2001).

Discussion and summation

Occupational violence appears to be an endemic risk for long-haul truck drivers, although it occurs in distinct forms in different places (road rage, at freight-forwarding yards, theft-related, and from dissatisfied customers). These forms of occupational violence are largely non-overlapping, the perpetrators have distinct profiles, the causes are different, and the most appropriate preventive interventions differ. The road rage form of occupational violence is potentially the most severe, and may have lethal consequences for a range of other road users.

- The risk of 'external' occupational violence is probably best reduced through enhanced CPTED-based control measures at freight yards, greater attention to locks that secure cargo, and fitting of sophisticated tracking systems within heavy haulage vehicles (Mayhew, 2001a).
- Road rage is probably best dealt with via broader societal measures including through the education system, the family (where much of this behaviour is learnt), road licensing authorities and by more stringent police enforcement of the road rules.

- Violence at freight-forwarding yards is probably best dealt with through strengthening the protective strategies implemented by CEO/managers – with additional assistance by Inspectors from the local OSH authority. It may be possible to relieve the underlying economic and time pressures through more effective organization of loading/unloading schedules, ensuring time schedules between pick-up and delivery are not excessively tight, and the establishment of standardized minimum freight rates per kilometer/mile – all of which may reduce the excessive competition that fuels aggression and gross overworking (Mayhew and Quinlan, 2000).
- 'Internal' occupational violence/bullying may be reduced by ensuring all those in the 'chain of responsibility' encourage equitable working environments, and meet their regulatory obligations.
- 'Client-initiated' violence is most likely to be effectively diffused through improved communication channels between service providers and customers – which recognize that long-haul drivers are usually only the 'messenger' in the middle.

That is, apart from the occupational violence associated with hold-ups (theft of goods or vehicle) and road rage, the prevention of occupational violence for long-haul heavy vehicle haulage drivers probably lies outside of the criminal justice system.

Courier drivers

Assessment of the occupational violence experiences of courier drivers is difficult as there are no substantive data and no known broad-ranging scientific studies were able to be located. Nevertheless, a general picture of the level of risk can be constructed from examination of the work of taxi drivers, pieced together with data from the limited research studies conducted, together with anecdotal reports.

Forms of occupational violence experienced by courier drivers
- Courier drivers may be at risk of 'external' violence as they occasionally carry precious cargo or highly sensitive documents. Such scenarios are not restricted to on-road deliveries with expensive cargoes occasionally carried on to aircraft and hand delivered to international customers. These cargoes may be legal or illicit (e.g. drug smuggling). Clearly the cargo carrier is at risk of a violent attack in such instances from competing criminals – or force during a police 'takedown'.

- Courier drivers have an increased risk of 'client-initiated' violence as their customers – by definition – require the goods in a hurry. Hence if delivery is delayed for any reason, customer annoyance can be expected. After all, the clients/customers pay a substantially increased delivery charge compared with official postage rates.
- Competition between courier drivers may lead to altercations about under-cutting when business volume is reduced.
- Finally, it has been assumed that on-road courier drivers are at high risk of road violence (road rage) as they spend much of their working time hurrying between pick-up and drop-off destinations.

Empirical evidence on experiences of occupational violence among courier drivers

There is limited evidence about the occupational violence experiences of couriers. However, 18 courier drivers based in Brisbane and Dalby (a small regional Australian town) were interviewed face-to-face and completed a semi-structured interview as part of a more comprehensive study into OSH in the transport industry (Mayhew, 1999b). Many courier drivers complained bitterly about the driving behaviours of other road users, discussed numerous crashes linked excessive working hours and fatigue, and cited various health complaints. Yet rarely was any time taken off work to recuperate from injuries or fatigue. As a local general practitioner who often treated injured self-employed workers stated in this study:

> Usually you've got to try and force them (self-employed) to stop work ... at the moment, economically; things seem to be very poor. There are a lot of businesses going to the wall. Quite a few who do see us would tend not to take time off. They would work on in spite of their injury.

However, in this small sample of courier drivers, no mention was made of 'external' hold-up related occupational violence.

Discussion and summation

Courier drivers appeared to be most at risk of the road rage form of occupational violence. The consequences of exposure to this aggression – together with long working hours and fatigue – may contribute to a cumulative stress burden and a poor overall OSH status (see: discussions in the second half of Chapter 8).

Maritime industry

There are diverse work environment factors that influence levels of exposure to occupational violence by maritime workers. Underlying labour process factors that may influence the level of risk include:

- Seafarers are usually employed on short-term contracts, work long shifts, live in cramped quarters, and are away from home base for extended periods of time (Mayhew, 2002c).
- The officers and crew are frequently of different nationalities. Crews are typically made up of a range of cultural groups who speak various languages, and who may have long-standing ethnic rivalries (Mayhew and Grewal, 2003).
- The legislative protections may be minimal for seafarers away from their home countries, at work on foreign ships, or in international waters (Clayton *et al*, 2000).
- Individual seafarers employed on 'flag of convenience' vessels may be very vulnerable, particularly those from developing countries (see: Couper, 1996). Certainly other working conditions are poor. For example, in a study of fatalities, Roberts (1998: 21) calculated that while 15% of British seafarers who died at work lost their lives through occupational injuries or maritime disasters, the figure was 40% for those on foreign-flagged ships.

Forms of occupational violence experienced by maritime workers

In the maritime industry, the 'external' form of occupational violence is usually manifest during piracy initiated for cargo or vessel theft. Maritime workers appear to be at increased risk of piracy when vessels are transiting high-risk geographical areas such as off the west coast of Africa, some South Eastern Asian waters and parts of the Caribbean (see: Mayhew, 2001a; Lane 1999; Couper, 1996).

The 'client-initiated' form of occupational violence is most likely to occur when a customer is dissatisfied with a service provided, for example, a passenger may attack a maritime worker enforcing bar closure times, or delayed delivery of goods may precipitate aggression from clients.

Seafarers are at risk of 'internal' occupational violence from co-workers, including verbal abuse, bullying, threats, assaults, gross initiation rites or sexual attacks. At the heart of many of these incidents is inadequate training, poor supervision, inappropriate management policies and strategies, or a culture that supports such behaviours.

Tensions may also rise because of shift work or overwork, language or cultural differences, or through the deliberate hiring of workers from a range of countries to ensure a non-cohesive workforce (Couper, 1996; Parker *et al*, 1997). The internationally acclaimed Australian *Ships of Shame* report explored a number of these issues. It uncovered extensive maltreatment of crew members including the denial of food and the provision of inadequate rations, the bashing of crew members by ships officers, the maintenance of two pay books (one for official records citing International Transport Federation (ITF) levels of pay, the other for the lower actual level of pay), inadequate accommodation and washing facilities, sexual molestation and rape, and deprivation of access to appropriate medical care (HRSCTCI, 1992: 36). While most of these incidents have reportedly occurred in international waters, a few have occurred within Australian jurisdictions (Mayhew 2002c). Isolated Australian incidents include the cases of Captain G. Griffin who was sentenced to nine months in jail in 1998 for assaulting a deckhand; and the disappearance at sea of Wayne Tyson in August 2000 after assaulting his captain (Taylor, 2000). On the *Glory Cape* (a Panamanian flagged vessel manned by Korean officers and Indonesian crew) in Australian waters in 1995, the Korean officers mounted a sustained beating of the Indonesian crew with iron bars. To escape these beatings, many jumped overboard and one crew member drowned. Remarkably, even though the dead seafarer had been extensively beaten, a post mortem report stated that there were no lesions or injuries to his body (see: overview in Clayton *et al*, 2000). Bullying between seafarers while vessels are at sea is of particular concern because offender and recipient(s) are constrained together for the duration of the voyage – unless violence becomes so severe that an emergency port call is required. Important intersections may exist between perpetrators and victims on the basis of hierarchy/seniority of position, ethnicity, tenured or precarious employment, power relationships between separate groups of workers on board ship, knowledge about employment rights, and organizational policies and strategies directed to the prevention of occupational violence and bullying (Mayhew and Grewal, 2003; Mayhew, 2002c).

While all forms of violence can affect maritime workers, it is probable that 'external' is least frequent but most severe (i.e. from pirates), and 'internal' (i.e. bullying from other workers on board) is most common (see: Couper, 1996; HRSCTCI, 1992). The internationally recognized violence researcher Duncan Chappell (2000: 299) stated unequivocally that: 'My inquiries concerning patterns and trends in

workplace violence associated with maritime transport failed to discover any direct or documented information on the subjects'. Thus,
apart from 'external' violence (i.e. piracy), the scientific evidence on
occupational violence among the international maritime labour force
is very limited. The remainder of this chapter focuses on estimation
of the incidence and severity of 'internal' violence experienced by
maritime workers.

Empirical evidence on occupational violence between maritime workers

A research study was initiated to provide baseline figures on the incidence and severity of occupational violence *between* seafarers. The aims
included:

- Estimations of the *incidence* of occupational violence/bullying;
- Estimations of the *severity* of occupational violence, including:
  verbal abuse, threats, physical assaults, initiation rites, sexual abuse
  and/or mobbing; and
- Identification of any variations between sub-groups of respondents.

A total of 108 students enrolled in a range of formal study programs for
the maritime industry participated in a face-to-face interview. The
majority were mature adults with nearly 70% concurrently working
full-time in the industry (Mayhew and Grewal, 2003). All interviews/
questionnaires were conducted face-to-face at the Australian Maritime
College (n = 71), the Kalmar University in Sweden (n = 23), and the
Malaysia Maritime Academy (n = 14). Each seafaring student interviewed was requested to provide information about any personal
experiences of occupational violence/bullying in the immediately previous 12-month period. A detailed semi-structured interview and questionnaire schedule was devised (based on that used by Mayhew in a
series of previous occupational violence studies). Adaptation of this
previous research instrument meant that the incidence and severity of
the different types of occupational violence experienced by the
maritime workers could later be compared with that identified for
some land-based jobs (see: discussions in second half of Chapter 8).

The pilot study found that the international seafaring labour force
exhibited all the core characteristics of a precarious labour force
(Mayhew and Grewal, 2003). Their labour market position was weak,
hours of labour were long, pay was comparatively poor, and occupational violence in the form of bullying was common. The data identified
a wide range of situations where 'internal' forms of occupational

violence occurred, although there may have been a slight tendency for the risks to be increased on 'Flag of convenience' ships. While there were significant demographic differences between seafarers interviewed in the three different countries, experiences of occupational violence were relatively consistent.

Of all 108 seafarers interviewed, 19.4% had been verbally abused at work in the immediately previous 12-month period, 5.5% had been threatened, 1% assaulted, and 1% experienced a sexual assault. Thus an annual incidence ratio of around 26.8% for all forms of occupational violence was recommended as a baseline ratio (Mayhew and Grewal, 2003). Cited incidents included: 'Verbally abused and threatened by shipmate'; 'Junior officers from Masters and Superintendents'; and 'Manager threatened an employee that if he didn't sign his workplace agreement on the spot he could leave. No time to read info'.

However, the *severity* of occupational violence/bullying experiences varied somewhat across seafarers from different countries. The Swedish workers experience more lower-level violence/bullying. In contrast, those interviewed in Australia and Malaysia tended to experience more severe incidents, including a male-on-male sexual assault (Mayhew and Grewal, 2003).

Each seafarer interviewed was also requested to provide information about incidents involving *other seafarers* that they had witnessed in the immediately previous 12-months. Fully 30.6% of the interviewees (again predominantly from Australia and Malaysia) had witnessed the victimization of other seafarers, including: '... radio officer was beaten up by the Master, resulting in bleeding'; and 'Abuse between races'.

Most aggression was 'top down' from more senior to subordinate crew members. Less commonly, subordinate seafarers became abusive to officers who imposed less favourable shifts or demanded excessive productivity. The second engineer on vessels appeared to be more commonly disliked and was vulnerable in already high-risk scenarios (Mayhew and Grewal, 2003). For example, 'Cook stabbed 2nd engineer. Drunk incident where some crew were distressed ...'. There were numerous reports that long hours of labour, shiftwork and increased tensions between crew members contributed to scenarios where occupational violence could erupt. As has been well-documented elsewhere, fatigue, stress and loneliness can cumulate and result in depressive illnesses – which in turn may fuel propensities to aggression (Parker *et al*, 1997). Yet some long-term seafarers had become inured to the risks of violence and regarded tense inter-personal relationships as the norm: 'The usual, in my last 10 years at sea'.

Discussion and summation

The pilot research study in the maritime industry provided baseline information about 'internal' occupational violence/bullying, including information on perpetrator characteristics and higher-risk scenarios. The international seafarers interviewed were in a comparatively poor labour market position, working hours were long, they were usually far away from their home bases, and came from a very wide range of ethnic backgrounds. Thus these workers were very vulnerable to a range of exploitative mechanisms, and experienced a significant level of lower-level 'internal' occupational violence/bullying.

The incidence of all forms of occupational violence can be reduced when *patterns* of aggression in the maritime industry are known, the risks are recognized and preventive interventions implemented. The risk of 'external' violence is most readily reduced through changes to the physical working environment and avoidance of high-risk geographical areas wherever possible (such as the Straits of Malacca). Effective prevention of 'internal' violence/bullying requires a stringently enforced zero tolerance of violence policy. Since both the International Maritime Organisation (IMO) and the ILO are related United Nations agencies, the *Code of Practice on Workplace Violence in Services Sectors and Measures to Combat This Phenomenon* may well be a core initial step in reducing the risks of 'internal' violence/bullying among the international seafaring labour force.

Conclusion

It is clear that transport workers are at risk of a range of forms of occupational violence. However, the incidence and severity varies between transport sub-sectors, reflecting variations in exposure to risk factors. While the extant data are inadequate, the relative risks for different forms of occupational violence can begin to be reliably estimated. Some of this early data appear in the table below.

Table 7.2 Estimated incidence of different forms of occupational violence by transport sub-groups

	Verbal abuse	Threats	Bullying	Assaults	Road rage	Other
Taxi drivers	81%	17%	–	10%	–	–
Heavy vehicle haulage	33%	8%	–	1%	21%	–
Maritime	19%	5.5%	–	2%	–	–

Overall, the data indicate that transport workers face a significant risk of lower-level occupational violence (e.g. verbal abuse), although more severe forms are only common in the taxi industry. Since these data were gathered through quite distinct research studies that were not formally linked, precise estimations of incidence and severity ratios remain equivocal. For example, the study of maritime workers was focused specifically on 'internal' forms of occupational violence/bullying. Nonetheless, the pattern of endemic victimization via verbal abuse is cause for concern.

There are a number of regulatory instruments that may assist with the prevention of all forms of occupational violence, including the criminal code, OSH requirements, and industry-specific requirements, for example, responsibilities placed on masters and owners of vessels. Regulatory obligations on employers, employees and the self-employed provide statutory impetus and enhance prevention efforts.

Under the Robens-based regulatory framework that provides the basis for the British OSH legislative framework and which forms the basis for that in place in all Australian states and territories, obligation holders are required under their 'duty of care' to provide a safe *place* as well as a safe *process* of conducting work. These obligations have been spelt out for self employed/atypical workers (such as taxi drivers) under some regulatory frameworks. For example, in the Australian state of Queensland the *Act* states:

> (1) A person (the 'relevant person') who conducts a business or undertaking has an obligation to ensure the workplace health and safety of each person who performs a work activity for the purpose of the business or undertaking. (2) The obligation applies: (a) whether or not the relevant person conducts the business or undertaking as an employer of self-employed persons; and (b) whether or not the business or undertaking is conducted for gain or reward; and (c) whether or not a person who performs a work activity for the purposes of the business or undertaking works on a voluntary basis (s29A of the Workplace Health and Safety Act 1995).

This clause clarifies the taxi operator's obligation to ensure the health and safety of taxi drivers (and similar workers) while performing a work activity for the taxi operator's business.

While ideally such obligations will be spelt out in some detail in *Acts*, subordinate legislation can also enhance reduction of the risk factors for all types of occupational violence. Codes of Practice can also detail situations where bullying behaviours are likely to arise, enumerate potential liabilities, and provide an outline of organizational policies that may counteract 'internal' forms of occupational violence. For example, WorkSafe Western Australia (the Western Australian OSH authority) has produced the *Code of Practice: Workplace Violence* (1999) which provides practical guidance on reducing the risks.

We believe that the *Code of Practice on Workplace Violence in Services Sectors and Measures to Combat This Phenomenon* is an essential prerequisite to reduction of the risks for all forms of occupational violence experienced by workers in a range of transport sub-groups across the world, including for the international seafaring labour force.

8

Occupational Violence/Bullying in Education and Juvenile Justice, and Assessment of the Impact from these Events

Claire Mayhew

As was argued in earlier chapters in this book, the evidence indicates that both the *incidence* and the *severity* of occupational violence are increasing across industrialized countries, particularly for workers who have significant levels of face-to-face contact with their clients or customers. Education workers – and those who provide other forms of care to adolescents – are not immune to these risks. There are two key themes in this chapter that relate to the experiences of occupational violence/bullying among education sector workers:

(a) The primary focus is identification of the risks of occupational violence faced by those who work with adolescents, for example, while providing educational or other forms of care and support. The discussions begin with a brief review of the international research literature. Evidence is then presented from two of our recent research studies.

(b) The second core focus of this chapter is assessment of the extent of emotional stress/injury/impact following an occupational violence/bullying event. It has long been assumed that those who suffer a physical assault are more likely to be emotionally traumatized by the experience vis-a-vis those who are merely verbally abused or bullied. Until now, insufficient objective and quantifiable comparative data have been available. However, as we have now conducted a series of studies in different industries, comparative data are available (see: latter part of Chapter 4, and Chapters 6 and 7). In recent studies, we have been measuring the extent of emotional consequences and have identified some interesting features. Data will be presented from a series of our studies that support the hypothesis that the emotional/stress impact from occupational violence/bullying events is not necessarily correlated with the physical severity of the incident.

The risks faced by those who work with adolescents

Increased levels of violence directed to service sector workers have been reported in the health industry (see: Chapter 6), and in geographical areas where the local population exhibits greater proportions of those with higher-risk demographic features (i.e. younger, male, substance-affected, remote area, or lower socio-economic) (see: Chapter 4). This evidence is instructive for education sector workers, particularly those who work with older adolescents in different geographical regions. Aggression from students clearly falls within the 'client-initiated' form of occupational violence (as defined in Chapter 2 of this book).

A range of risk factors have been identified as frequently associated with aggression from adolescent students, including: (a) classroom environment factors (e.g. availability of alcohol and illicit substances, clarity of behavioural norms, consistency of rule enforcement, emotional support, and strength of leadership); and (b) individual characteristics, including level of attachment to school, performance, exposure to negative peers, low self-control, early problem behaviour, ability to access weapons, and attitudes favouring legal violation or drug use etc. (Gottfredson, 1998). Other environmental influences include juveniles who have both delinquent friends and 'problem' parents, exposure to family violence, gang membership, teenage promiscuity, dropping out of school, physical abuse or neglect in early youth, and early independence from family (Howell, 1995: 4–5). However, it is beyond the capacity of this chapter to detail all the risk factors for adolescent aggression. Rather, the discussions are focused on violence from adolescents that is directed to the workers who educate or care for juveniles.

Abuse of alcohol, use of illicit substances, and social disadvantage are all risk factors for offending and for recidivism among adolescent offenders. For example, in assessing the risk factors for recidivism, the authors of a US study argued: '... problems in school and family, as well as substance abuse, are consistent with ... those who reoffend and adjust poorly in the community ...' (Heilbrun *et al*, 2000: 288; see also Howell, 1995). Similarly, when the offending trajectories of 1,503 Australian juvenile offenders were tracked over a 7-year period it was found that 79% progressed to the adult correction systems, with the risks of recidivism increased among: males, those with high levels of instability in their lives, low literacy, Indigenous background, poor employment prospects, disadvantaged socio-economic status, and substantiated maltreatment (Lynch *et al*, 2003: 2–5). In addition, the Australian Drug Use Monitoring (DUMA) study identified high rates of recidivism and progression to the adult corrections systems for those who used illicit substances (Wei *et al*, 2003).

Among juvenile *detainees* a very high level of illicit substance use has been identified across a number of countries. The DUMA project is conducted quarterly with the aim of providing aggregated data on illicit drug use among police detainees. Of 360 juveniles aged between 11 and 17 years recently assessed, 73% had used one or other drug during their lifetime, 46% had used two or more, and 47% used drugs other than cannabis (Wei *et al*, 2003: 3). In the immediately previous 30 day period, 54% reported that they had used cannabis, amphetamines (16%), ecstasy (13%), illicit opiates (13%), cocaine (7%), illicit benzodiazepines (5%), hallucinogens (2%), and street methadone (1%) (ibid). These self-reported behaviours were backed up by data from urine screen tests which indicated that 48% had used cannabis, 12% opiates, and 11% amphetamine (ibid). Of particular relevance for educators, the average age of initiation for cannabis was 13 and 14 to 15 for other illicit substances, and '… juveniles who had recently used heroin, cocaine or amphetamines self-reported much higher levels of offending …' (Wei *et al*, 2003: 5). Thus, a range of risk factors are likely to contribute to juvenile offending, many of which will be identifiable by the educators who deal with adolescents on a day-to-day basis.

Many detained juveniles have committed violent offences. While the Australian juvenile detainees were often charged with more than one crime, violent offences (22%) were the second highest category after property offences (54%) (Many had used illicit substances) (Wei *et al*, 2003: 3). Similarly, in a study involving 140 male juveniles conducted in the US state of Virginia, 41% had committed violent crimes (most commonly assaults), 43% property crimes, and 13% drug offences (Heilbrun *et al*, 2000: 279; see also Howell, 1995).

In the USA, a series of incidents perpetrated by students wielding firearms in schools and universities have been reported in the media. Two recent serious incidents at Australian Universities provide dramatic evidence of the increased level of risk to educators in that country (although still less common than in the USA). The first warning came from La Trobe University when a sacked student worker attacked on-campus bar staff. A second fatal incident at Monash University in late 2002 involved shots being fired at both staff and other students. According to newspaper reports, in both these incidents the student perpetrators were under significant stress. Together, these fatal and near-fatal incidents provide clear warning signs for educators, particularly those who deal with older adolescents or teenagers who are members of higher-risk categories.

We have conducted two empirical studies in the previous 18 months in Australia that are of direct relevance, as well as a series of studies in other industry sectors that provide comparative data and indicate some possible preventive interventions and strategies.

Methodologies adopted in recent research studies

Each of the studies discussed below was designed within an inductive approach. With inductive research, theory is built from the ground up following the gathering of extensive empirical evidence (Neuman, 1994: 41). Such studies identify basic problem characteristics, may indicate cause/effect relationships, provide baseline data, and indicate directions for formulation of hypotheses for testing in later studies. For example, in inductive studies, violence may be identified as commonly associated with perpetrators having a particular characteristic, or in explicit high-risk scenarios. As with more traditional study designs, in inductive research temporal order logic requires that the cause must come before the effect, associations between variables be identifiable, and potential causal relationship highlighted (Neuman, 1994: 45). Nevertheless, inductive researchers would normally accept that field conditions, the contexts of events, mutually influencing variables, emotional intensities and resistance all have the potential to influence the severity of occupational violence/bullying events.

Inductive research methodologies are essentially linked with what have become known as interpretive social science methodologies and ideas, including hermeneutics (theories of meaning), ethnography (cultural understandings), phenomenological and symbolic interactionist (subject and researcher influence each other in a dialectical fashion) ideas. All of these methodologies adopt participant observation or field research techniques involving close contact between researcher and subject so that the social reality under study is understood from the standpoint of the participants, usually within a practical orientation.

Each of the empirical studies we have conducted (either together or separately) in the different Australian industry sectors and which are discussed below were established independently, conducted in distinct places, and had a slightly different focus. However, the primary researcher was the same in each of the studies and she developed, adapted and applied a similar semi-structured questionnaire instrument in all research projects and also adopted similar methodologies. Hence the findings allow for direct comparisons.

Background to studies and sample selection

(a) *Juvenile justice:* Our most recent study was focused on workers in a juvenile detention centre (Mayhew and McCarthy, 2003). The different occupational classifications employed in the broader juvenile justice workforce include: youth workers, social workers, community corrections/parole staff, psychologists, administration, health professionals, educators, and support staff etc. Within Australia, these workers cared for an estimated 491 male and 54 female juvenile detainees in 2002, as well as many more in the community who were subject to particular orders (Bareja and Charlton, 2003: 11–12). Indigenous people and young males were notably over-represented (ibid, 2003).

In order to provide baseline data on experiences of occupational violence/bullying among the workers who care for these detained juveniles, we conducted a pilot study in late 2003. Fifty workers in a (nameless) juvenile detention centre within Australasia were interviewed face-to-face and completed a semi-structured interview schedule that required both qualitative and quantitative responses to a series of questions (Mayhew and McCarthy, 2003). Interviewees were selected through an allocated number selection process (every fourth person on a total workforce employment record compiled in alphabetical order). The resulting interviewee population included males (64%) and females (36%), a range of age groups, and workers employed on a full-time (52%), part-time or short-term contract (8%), casual (34%) or some 'other' basis including a few who provided specialist services to the young people detained (6%) (Mayhew and McCarthy, 2003). Of these 50 interviewees, the majority (60%) worked rotating shifts. Fully 74% belonged to a trade union (which may therefore provide a useful medium through which information about violence prevention strategies can be disseminated).

The core aims of this pilot research study were to establish baseline estimates of occupational violence/bullying experienced by the juvenile justice workforce and to identify the risk factors. Data collected included the *number* of violent events experienced over the previous 12-month period, witnessing of events directed to other staff members, alleged perpetrator characteristics, perceptions of higher-risk locations and situations, and suggested prevention strategies etc. The *severity* of each event was estimated by separating events into verbal abuse, bullying, threats, initiation rites, assaults, or 'other' activities such as spitting at staff members. Among other things, our aims were to gather sufficient data to enable estimations of:

- Incidence and severity;
- Variations in patterns between the different occupational groups;
- Characteristics of perpetrators;
- Identification of 'hot spots';
- Detection of perceived key risk factors; and
- Development of preliminary policy recommendations to reduce the risks faced by juvenile justice workers working in youth detention centres.

(b) *Tertiary education:* The previous research project focused on identification of the violence/bullying risks from students in tertiary education. In 2002/03, 100 staff members of a multi-campus Australian university were interviewed face-to-face and completed a semi-structured interview schedule. An important caveat was that this was a *self-selected* population who had responded to an invitation for an interview during an electronic survey of the whole workforce (see: McCarthy *et al*, 2003b). The categories of data collected were similar to those adopted in the juvenile detention worker study discussed above.

The data from other research studies conducted in different industry sectors are used for comparative purposes in the discussions later in this chapter. The methodology adopted in the cited *health violence* study was detailed in Chapter 6 of this book. For both *seafarers* and *long-haul transport* drivers, the methodologies are described in Chapter 7.

Occupational violence/bullying experiences of juvenile justice and tertiary education workers

Juvenile detention

The 50 staff members interviewed had experienced an average 2.2 incidents each over the previous 12-month period. That is, 76% had personally experienced 109 separate incidents where they had been verbally abused, bullied, or assaulted. Behaviours experienced during these 109 separate events included: verbal abuse (67.9%), threats (35.8%), assaults (17.4%), bullying (11.9%), or some other form of aggression, such as being spat on (12.8%). (Because many perpetrators adopted multiple tactics during violent events, the total exceeds 100%, for example, both threats and an assault.) In marked contrast to the experiences of tertiary education workers discussed below, other staff members perpetrated only 16.5% of the 109 incidents and the detained juveniles perpetrated 83.5% of the described events. Those most at risk of 'client-initiated' violence in this centre were staff with close face-to-face contact with the detained juveniles.

> ... basically our client group on workers/employees. Verbal abuse continually, every shift. Threats happen all the time e.g. 'going to kill you; kill your family; rape your wife; kill your children' to that degree mostly. Assaults – lost count, about four to five I've witnessed. A common thread is these young people are in crisis with the law, the governance they are held under, and that they have on-going issues with their own families and themselves.

Those staff members who had not experienced any events in the previous 12-month period were overwhelmingly rostered in areas with limited face-to-face contact with the detained juvenile, for example, in kitchen areas or working on administration duties. 'Internal' co-worker aggression was minimal, but occurred across a spectrum of job tasks.

Eight *core risk factors* for adolescent aggression against juvenile detention workers were indicated:

- Juveniles who had been convicted for violent behaviours;
- Juveniles who had experienced physical or sexual abuse in their home environments;
- Juveniles engaged in altercations with other adolescent detainees;
- Staff directions to juveniles, including orders to complete minor personal tasks;
- Juveniles with poor impulse control;
- Membership of a 'gang' that was in conflict with another group of young people detained within the same juvenile justice centre;
- Juveniles who *perceived* that they were receiving discriminatory treatment based on their race/ethnicity; and
- High-stress precursor risk factors for violence by juveniles included negative experiences with family and friends (e.g. some phone calls).

Typical scenarios and pre-cursor situations described by the juvenile justice workers interviewed included:

'... the kids physically abused by their parents, the only thing they know is to use violence ...'.

'You tell the young person to do something and they just ignore you, swear, tell you to f... off. Eventually they do what you tell them to do, but they draw it out'.

'He had a bad phone call and just went off chucking chairs and tables around.'

'... the ones who are in for violent crime. They are criminals and people get so surprised when they display criminal behaviour.'

'... I walked in the door and ... hit me in the head with a chair ...'

'... Basically he walked up to me and whack – no words, no warning, no reason, nothing ...'

'... six years ago mainly cars ... gone from car theft to more grievous bodily harm and assaults; a lot of rape things now', and

'... the seriousness of the crimes is increasing since about mid 1995 onwards e.g. serious assaults. Also increasing is the period of time on remand which must affect the kids ... if no clear light at end of tunnel, the anxiety and fear add to propensity to violence' .

We concluded from the pilot study data that client-initiated aggression is an endemic risk for juvenile detention workers who have face-to-face contact with their clientele. Of note, the severity of these aggressive acts was considerably higher for workers in juvenile detention than in other industry sectors where we have completed studies (see: below).

Tertiary education

Of 100 University staff members interviewed, 80% had personally experienced 99 separate incidents where they had been verbally abused, bullied, or assaulted in the previous 12-month period (McCarthy *et al*, 2003). The events included: treated unfairly (65%), denigration (58%), verbal abuse (50%), unreasonable work practices

(47%), threat (39%), offensive words (30%), assault (1%), and 'other' (25%) (ibid). (Many experienced more than one tactic in the same incident, so totals exceed 100%.) In addition, 74% had witnessed (or overheard) 90 separate events where other staff members were being victimized over the same period. However, students perpetrated only 13% of these incidents. Hence we concluded that client-initiated aggression was relatively uncommon vis-à-vis reports of bullying, unreasonable work practices or other forms of aggression perpetrated by other staff members (Mayhew *et al*, 2003).

Nine *core risk factors* for student aggression in tertiary institutions were identified:

- Potential failure of assessment item or subject;
- Accusations of plagiarism;
- Significant financial stress;
- Requirement for field placement when child care or travel was difficult;
- Assessment on personal capacities when unclear marking criteria provided;
- Culturally challenging tasks and relationships;
- Work or family commitments that conflicted with University timetables;
- Angry or stressed students, particularly when dealing with under-resourced staff; and
- Students with a mental health or alcohol/substance abuse problem (Mayhew *et al*, 2003).

All these risk factors may exist in other educational facilities to a greater or lesser extent, particularly where the clientele are predominantly older adolescents.

Comparisons across industry sectors

Because reliable data on a range of forms of occupational violence are scarce, we have compared our empirical data sets across a series of studies that we have been involved with in recent years. In total 1,264 individuals have been interviewed face-to-face and completed a similar semi-structured interview schedule that requires both qualitative and quantitative responses to a series of questions about occupational violence/bullying (see: detailed descriptions above). In each study, interviewees were requested to supply precise information about all violence/bullying incidents experienced over the immediately previous 12-month period, alleged perpetrators, perceptions of higher-risk locations and situations, and suggested strategies for prevention. The 12-month time-span was selected to assist with calculation of an incidence ratio. In each study, the *severity* of each violent/bullying incident was estimated by separation of perpetrator behaviours into verbal abuse, bullying, threats, initiation rites, assaults, or 'other' activities (e.g. spitting) (Mayhew, 2000b).

Some of these comparative data appear in the table below. In this table it is important to note that more than one type of behaviour was used in many of the violent incidents; hence row totals can exceed 100%.

The data presented in Table 8.1 indicate that:

- The vast majority of cited incidents involved verbal abuse or other *non-physical* threats to the well-being of workers.
- Verbal abuse and threats are common experiences in many jobs.
- Exposure to assaults was highest among the juvenile detention and health workers.
- Bullying appeared to be more commonly experienced by tertiary education workers; all of whom had responded to a request for interview during an electronic survey focused predominantly on bullying (i.e. they were a self-selected sample).

Table 8.1 Forms of occupational violence experienced over 12-month period, by industry sector (% of interviewees)

	Verbal abuse	Threats	Assault	Bullying	Other	Total victimized
Juvenile justice (n = 50)	67.8	35.7	17.4	11.9	12.8	76
Tertiary ed (n = 100)[1]*	50	39	1	65*	25	80
Health care (n = 400)[2]	67	32.7	12	10.5	11	67.2
Seafaring (n = 108)[3]	19.4	5.5	1	**	1	26.8
Long-haul transport (n = 300)[4]	32.7	7.7	0.7	**	21***	
Fast-food (n = 304)[5]	48.4	7.6	1	*	2.3	

Notes: In each of the studies 'bullying' was defined by each individual interviewed.
* The tertiary education workers were a self-selected sample who had probably been disproportionately victimized through bullying. The 65% 'bullied includes those who cited treated unfairly (65%), denigration (58%), and unreasonable work practices (47%). ** Bullying could not be separated out in the seafaring or transport studies. *** The 'other' category was predominantly road violence/road rage.
References: (1) McCarthy *et al*, 2003b & Mayhew *et al*, 2003; (2) Mayhew & Chappell, 2003; (3) Mayhew and Grewal, 2003); (4) Mayhew and Quinlan, 2001; (5) Mayhew, 1999c.

- However, it is important to note that there are categorical overlaps between occupational violence and bullying (e.g. verbal abuse). In the studies described in this Chapter, interviewees decided upon the category of occupational violence/bullying for themselves.

While the data in Table 8.1 provide clear indicators of the *extent* of different forms of occupational violence/bullying in different jobs, this says little about the relative *impact* on the recipients. Until now, it has been difficult (if not impossible) to compare the emotional stress impacts on recipients/victims following victimization through different forms of occupational violence.

The final part of this chapter focuses on: (a) the utility of one instrument, the General Health Questionnaire (GHQ-12), that can be used to measure the emotional stress/injury/trauma consequences of occupational violence/bullying and the application of this in a series of large-scale empirical studies conducted in the juvenile justice, tertiary education, health, seafaring, and long-haul transport industries; and (b) an evaluation of the extent of emotional stress/injury/trauma experienced by the 958 workers interviewed face-to-face during these studies.

Measurement of the negative health effects from occupational violence/bullying at work

A series of instruments have been used to measure the emotional stress and psychosomatic effects following exposure to aggression at work. The severity of *emotional stress* repercussions from occupational violence/bullying can be assessed through use of the General Health Questionnaire (GHQ). The GHQ is an internationally validated instrument that measures the *current perceived* state of stress and predicts future health status. Because it has proved to be reliable, the GHQ is widely used in medical and psychosocial studies and has been repeatedly validated in Australian and international studies (Chapman, 2001; Goldberg *et al*, 1997; Goldberg and Williams, 1991; Graetz, 1991; Banks *et al*, 1980; Goldberg, 1972). Nevertheless, use of the GHQ is rare in empirical studies, particularly in occupational safety and health (OSH) and criminological studies of occupational violence.

The GHQ was originally developed as a 60-item questionnaire (GHQ-60), but 30, 28 and 12-item versions were subsequently developed. The GHQ-12 is the shortest version, yet retains the high validity and is particularly attractive to those working in busy services industry clinical settings where there are time constraints and interviewees have multiple demands on their time (Goldberg *et al*, 1997: 191). All versions of the GHQ have pre-set questions that have numerical scores allocated for each response, which are subsequently totalled to give an overall score. The GHQ-12 can be scored via a Likert scale that has a potential range from 0 (substantial decrease in all symptoms) to 36 (substantial increase in all symptoms) (Goldberg and Williams, 1991: 19, 63; Graetz, 1991: 133; Goldberg, 1972: 36, 57).

Past studies have indicated that, using the simple Likert scaling method, a GHQ-12 score of between 8 to 10 is relatively normal with a threshold of around 11 or 12; a person with a score greater than 14 probably requires urgent assistance (Goldberg and Williams, 1991; Goldberg, 1972: 91). The threshold score has been defined as when: '... the number of symptoms at which the probability that an individual will be thought to be a case exceeds 0.5' (Goldberg and Williams, 1991: 3).

Nevertheless, some difficulties have been reported in administering the GHQ to particular groups. Goldberg (1972: 82, 96–97) identified three groups whose poorer health status was likely to be missed: (a) those not prepared to describe symptoms in a 'pencil and paper' situation or who were illiterate; (b) people with dementia, chronic schizophrenia or hypomania; and (c) those with very longstanding disorders going through a latent phase of their illness, for example, post-traumatic stress disorder (PTSD) may remain undisclosed in the short-term (Chapman, 2001: 244; Goodchild and Duncan-Jones, 1985). That is, since the GHQ focuses on symptoms over the previous month, the delayed effects from assaults that eventually manifest as PTSD may not show up in the short-term. There are also some social confounders, for example, females usually report higher scores than do males, and while divorced and separated people tend to have higher mean scores, widows and widowers do not (Goldberg, 1972: 96–7). A further threat to validity is that interviewees may record 'donkey' responses through checking the same column in response to separate questions on the GHQ. Nevertheless, overall, the GHQ remains a robust objective instrument (Goldberg and Williams, 1991: 74; McPherson and Hall, 1983; Banks *et al*, 1980: 193; Goldberg, 1972: 82–99).

It is also debateable whether the GHQ-12 is suitable for both occupational violence and bullying. Occupational violence is concerned more with the one-off incident and less with the repetition of less overt events. Thus most occupational violence researchers would consider the GHQ a current status-quo measure of distress. In contrast, if a researcher was setting out to measure the traumatic impacts of bullying, the GHQ would probably not be the instrument of choice for many. They would most likely use an abbreviated PTSD scale, or a battery of scales, that measured how repeated re-experiencing of events, places and people factored into the level of distress and trauma. Nevertheless, the author believes that the GHQ-12 is the best available instrument to measure the impact of both overt occupational violence and covert bullying.

The GHQ-12 form was attached to the semi-structured interview questionnaire used in each of our recent research studies in the juvenile justice, tertiary education, and health, seafaring and long-haul transport studies, in order to estimate and compare the emotional stress impact following a violent/bullying event. Because a Likert-scaled GHQ-12 was used in each study, the emotional stress/repercussions from occupational violence/bullying could also be compared across these industry sectors.

Research evidence: GHQ-12 scores following exposure to occupational violence/bullying

During the course of interviewing the 958 workers employed in the juvenile justice, tertiary education, health, seafaring, and long-haul transport industries, extensive emotional stress/injury/trauma consequences from occupational violence and bullying were identified (with a caveat that the tertiary education interviewees were a self-selected population). The distinct empirical studies provide contextual evidence about the nuances of impact associated with different forms of occupational violence/bullying.

The data in Table 8.2 indicate:

- A clear rising trend in increasing GHQ-12 score (i.e. emotional stress/injury) that is closely correlated with increasing exposure to occupational violence/bullying events in the immediately previous 12-month period.
- The GHQ-12 impact is, proportionately, higher for those who have experienced bullying. In many cases, covert forms of denigration resulted in more extensive emotional trauma than did physical abuse.
- That is, the physical severity of an incident is not necessarily correlated with the extent of emotional impact.
- While the data in Tables 8.1 and 8.2 indicated that juvenile justice workers experienced a higher incidence of more *severe* violent events

Table 8.2 GHQ-12 scores following occupational violence/bullying experiences

	Nil	1 event	2 events	3 or more	bullying (separated out)	overall
Juvenile justice	10.0	8.13	11.67	13.09	16.0	11.36
Tertiary education[1] *	9.24	14.37	13.82	15.12	16.16	13.6.
Health industry[2]	9.98	10.99	12.45	12.65	15.22	11.42
Seafaring[3]	11.3	13.81	14.0	–	**	11.84
Long haul transport[4]				11.87	**	10.3

(*Notes*: * the tertiary education workers were a self-selected sample who had probably been disproportionately victimized through bullying. ** Bullying could not be separated out in the seafaring or transport studies.)
References: (1) McCarthy *et al*, 2003b & Mayhew *et al*, 2003; (2) Mayhew & Chappell, 2003; (3) Mayhew and Grewal, 2003); (4) Mayhew and Quinlan, 2001.

compared with workers in the other industry sectors we have studied, it appears that workers do not build up a resistance to the emotional/stress consequences. Rather, as noted in Chapter 6, it is clear that workers do not become inured to verbal abuse, threats or assaults at work (Mayhew and Chappell, 2003).

Some industry-specific findings are detailed below.

Juvenile justice

There was a clear rising trend in GHQ-12 scores that was correlated with increased number of violent events experienced in the immediately previous 12-month period. While bullying from colleagues was rare, this type of aggression had a significant impact.

> ... Belittling, intimidation, raising of his voice, and just plain old ignoring me. Happened any interaction we had. A great sense of power on his part ... when you are put into a position of powerlessness, mediation via HR management doesn't seem a viable option.

It may be that some of this inappropriate behaviour is unwittingly facilitated by stark divisions between different groups working in the detention centre: (a) tensions between a 'corrections' versus 'rehabilitation' focus; (b) a male dominated workforce; (c) different ethnic groups; and (d) a tendency for the workforce to polarize between long-term employees and more recent casual workers.

Tertiary education

The *impact* on the recipients was significant. Recipients of aggression had an overall score of 14.71, witnesses averaged 13.95, and those who neither witnessed nor experienced incidents had an overall score of 10.0 (McCarthy *et al*, 2003b). Experiencing bullying/violence incidents was correlated with a GHQ score that was above the clinically significant level of 14.0, closely followed by being a witness to such events. Fears of future further incidents also escalated the level of trauma. Nevertheless, because this was a self-selected sample, the findings for recipients must be treated with caution.

Health industry

There was a steady rise in GHQ-12 score correlated with increased expose to violent events, which was significant at the 0.0001 level (Mayhew and Chappell, 2003). Unexpectedly, the relationship between exposure *to assaults* and GHQ score was not statistically significant (0.2198) (ibid). Those subjected to bullying from other staff members had the highest GHQ scores of all, which was significant at the 0.0001 level (ibid). The authors concluded that physical severity and impact were not necessarily correlated with the level of emotional stress impact (Mayhew and Chappell, 2003).

Seafarers

There was a steady rise in GHQ-12 score correlated with increasing exposure to occupational violence/bullying. Long hours of labour, shift-work, fatigue and stress were reported to result in increased tensions between crew members and contribute to aggression: '... abuse and assaults against crew' (Mayhew and Grewal, 2003). Because most sea-farers usually work far from their home support networks, their vulner-ability to a range of exploitative mechanisms was heightened (Mayhew, 2002c). Further, seafaring crews are often multicultural, and this some-times exacerbated tensions, for example, '... abuse between races'. Those working on 'flag of convenience' ships appeared to be at greater risk. However, some long-term seafarers appeared to have superficially become inured to the risks of violence and regarded tense interpersonal relationships as the norm: 'The usual, in my last 10 years at sea'.

Long-haul transport

A number of work-related variables contributed to high GHQ-12 scores, *in addition to the emotional trauma induced by occupational violence/*

bullying. For example, scores varied by employment status with owner/drivers having a mean GHQ-12 of 11.5, compared with small fleet drivers (9.8), and those working for large fleets (10) (Mayhew and Quinlan, 2000).

High GHQ-12 scoring workers

As noted earlier, a GHQ-12 score of 14 or above is considered to indicate that a worker is in significant distress and probably requires urgent professional assistance. Hence we separated out high-scoring workers from each of the studies.

As can been seen from the data in Table 8.3, a concerning proportion of the workforce in each of the industries scored at or above the clinically significant level of 14 (see: Goldberg *et al*, 1997; Goldberg and Williams, 1991; Goldberg, 1972). Among both the juvenile detention and the tertiary education institution, a significant proportion of the workers scored at very high levels. While the tertiary education scores may have been artificially raised through the self-selected process (i.e. more victims are likely to have volunteered to participate), the juvenile justice workers were selected by a sequential allocated number system. That is, the juvenile justice workers interviewed (and their scores) were likely to be representative of this workforce as a whole.

The high level of overall GHQ-12 score by juvenile detention centre workers suggest that this workforce operates under a significant degree of stress. Assuming the 'healthy worker' effect is in operation – where those unable to tolerate occupational violence leave and only the most 'hardy' remain – the emotional/stress consequences of violent events are even more of a problem than the data from current workers indicate.

Table 8.3 **High GHQ-12 scores from juvenile justice, tertiary education, health, seafaring, and long-haul transport (using Likert scaling)**

	Overall score	*% with GHQ-12 score 14 or above*
Juvenile justice (n = 50)	11.36	42%
Tertiary education (n = 100)[1]	13.6	42%
Health (n = 400)[2]	11.42	22.5%
Seafaring (n = 108)[3]	11.84	30%
Long-haul transport (n = 300)[4]	10.3	15.7%

References: (1) McCarthy *et al*, 2003b & Mayhew *et al*, 2003; (2) Mayhew & Chappell, 2003; (3) Mayhew and Grewal, 2003); (4) Mayhew and Quinlan, 2000.

Possible prevention strategies

We identified some protective strategies that reduced the risks of occupational violence in the juvenile detention centre, which may fruitfully be applied to other settings where adolescents are educated, treated or cared for. However, it is of core importance to note that occupational violence was *expected* from adolescents in custody, the facility had been designed to reduce the risks, and staff members were provided with emergency equipment and training to deal with juvenile aggression. Another positive feature identified in the juvenile detention centre was the rapidity of emergency back-up assistance. For this (and other) reasons, juvenile justice workers in the community (as well as others who work with adolescents in neighbourhood areas) may be at far greater risk than those based within juvenile detention centres. As one interviewee stated: 'For the people who work in the field, they don't have the back-up we have'. Our research findings back up this quotation, for example, in child and adolescent wards in hospitals our studies have revealed that facilities are generally poorly designed to reduce the risks of occupational violence, many CEO's consider client aggression an aberration, and staff are frequently inadequately prepared to deal with both low-level and overt violence from clients/customers/patients.

To assist readers who work with adolescents in designing and implementing an occupational violence/bullying risk reduction strategy we have briefly listed some basic preventive strategies:

- A *pre-planned, multi-faceted and organization-wide* approach needs to be adopted.
- Facilities should be designed with the assistance of experts in 'Crime prevention through environmental design' (CPTED). CPTED principles focus on the design/refurbishment of buildings and environments in such a way that violent crimes are more difficult to execute. Strategies include what is known as 'target hardening' (making it difficult for perpetrators to access parts of buildings or valuables), 'increased surveillance (the aim is to make perpetrators more visible in the hope that this may dissuade aggressive acts e.g. closed circuit television), and 'decreasing the rewards' (reducing the benefits from commission of a violent act) (see: discussions in Chapter 4; Crowe and Adams, 1995).
- An emphasis on consistency in application of (a) 'carrots' that reward good behaviour, and (b) 'sanctions' applied to juveniles who do not comply with staff expectations and requests.

- Enhanced preventive interventions and rehabilitation efforts among families and in facilities, for example: 'If we worked to get rid of the violence in the first place, the kids won't be charged again'.
- In the community, 'whole of government' approaches to juvenile violence and offending may involve a range of agencies, including housing (e.g. more stable home environments), education (e.g. improved literacy), health (e.g. Hep. B immunizations), local government (e.g. provision of swimming pools or sporting fields in disadvantaged areas), as well as: '... crucial importance of targeted early interventions that address the precursors to juvenile offending ...' (Lynch *et al*, 2003: 5).
- Education of the public at large about offences committed by detained adolescents, and the precursor risk factors for juvenile offending, e.g.: '... the people in head office need to come and work on the floor for a couple of weeks to get the feel for the actual job ... The community should be more supportive of the workers in here ...'.
- Clarification of the 'workplace culture' orientation for all staff members and juveniles. For example, the system can be clearly focused on rehabilitation of the young people, with a de-emphasis on the 'corrections' culture and sanctions.
- Clear application of procedural justice for workers accused of inappropriate behaviour by the detained juveniles, for example: 'Big issue in the department is taking allegations from young people who are convicted criminals and well-known liars as gospel over the youth workers'
- Development of a 'zero tolerance' of verbal abuse and workplace bullying between staff members that is *enforced*. '... You can't go to your boss and you can't go to your line manager. Are there any independent people in the workplace you can go to about a bullying problem?'
- Employment of the majority of staff on permanent contracts with a decreased proportion of casual staff
- Employment of a workforce with a range of cultural/ethnic backgrounds to match the clientele profile as far as is possible.
- Prohibition on the wearing of ties, scarves and jewellery by staff members (which may be used to attempt strangulation or other injury)
- Training and re-training of all members of the workforce in interpersonal skills, aggression minimization, governance rules,

consistency in rewards and sanctioning of juveniles, zero tolerance of bullying and sexual harassment (and what behaviours
constitute inappropriate workplace conduct), and emergency
planning, for example:

> ... need more training in dealing with violent behaviour ... and not
> from people who don't know the job we do, very important. Even
> people from adult corrections have no idea. It's a different client
> base ...

However, it is *not* sufficient to solely focus on violence training and
emergency responses, as these are not effective *prevention* strategies on
their own. Going for the cheap option (training to cope better with
aggressive behaviour) will not save money in the long run; rather a
false sense of security is likely to arise in the short-term – and larger
financial penalties in the longer term.

A series of recommendations were also made to reduce the risks of
adolescent student aggression, all of which can be considered by CEO's
of universities, technical colleges and secondary schools:

- Development of a student Code of Conduct
- Provision of detailed guidelines on what constitutes plagiarism
- Clear identification of the sanctions that apply to those who
  commit plagiarism
- Clear assessment criteria for all programs
- Comprehensive emergency planning
- The appointment of a University ombudsman
- Development of a mental health policy and strategies for both staff
  and students (Mayhew *et al*, 2003; McCarthy *et al*, 2003b).

The human resource impacts from 'client-initiated' violence/bullying
will inevitably increase over time in all industry sectors if prevention
strategies are not implemented, if the interventions are not tightly targeted, or if the incidence and severity of occupational violence multi-
plies. As detailed in chapter 3 of this book, the evidence indicates that
there are significant economic costs associated with covert and overt
forms of occupational violence and bullying (see also: Hoel *et al*, 2001).
Economically and socially, it is likely to be far more effective to address
occupational violence/bullying when these inappropriate behaviours
first arise, than it is to allow concerns to fester in a way that may
trigger extreme behavioural reactions.

Conclusion

All the indicators are that the incidence and severity of occupational violence is increasing over time. The best available evidence is that *at least* 10% of workers in industrialized countries will experience some form of occupational violence at work each year. Hence wise administrators will be prioritizing prevention strategies as there are extensive direct and indirect costs (see: Chapter 3). The 'emotional labour' component of job tasks may also need to be considered during analyses because the *intensity* of demands increase when workers are required to repeatedly deal with inappropriate behaviours such as verbal abuse (Boyd, 2002). Without remediation, the end result may be higher stress levels, anxiety, emotional exhaustion and burnout.

Employees in the tertiary education industry appear to be at particular risk of the 'client-initiated' form of occupational violence (see: Chapter 2), although substantive international evidence is patchy. Nine core risk factors that fuel student aggression in tertiary education were enumerated, including allegations of plagiarism, financial stress, child care needs, cultural differences in behaviours and expectations, time-demands because of concurrent employment, the demands from field placements, and specific study requirements.

The risks to staff providing care to adolescents in detention were also reviewed. The authors argued that the design of modern juvenile detention centres reduces the risks to a significant level. The design of such modern juvenile detention centres provides important guidance for the design of new secondary high schools in higher risk geographical areas. However, additional human resource employment and workplace 'cultural' efforts are also necessary to reduce the risks from juvenile justice clientele who are generally more volatile than in most other industry sectors.

Thus, our data from workers who routinely deal with adolescents are overwhelming: occupational violence must now be considered to be an *endemic risk* for all workers who educate or care for juveniles, particularly those dealing with teenagers under significant stress, from a disadvantaged socio-economic background and/or who abuse alcohol or use illicit substances.

The mounting evidence from a range of industry sectors clearly indicates that the impact from any form of occupational violence/bullying perpetrated by colleagues can, in many instances, equal the emotional trauma following assaults on-the-job and have a significant detrimental impact. Similarly, a study of 254 employees in 71 different occupations reported:

> The experience of co-worker aggression, but not the experience of public aggression or violence, directly predicted emotional well-being, psychosomatic well-being, and affective commitment to the organization (LeBlanc and Kelloway, 2002: 462).

Our recent research also indicates that the enduring nature of bullying and the resultant ongoing trauma may produce more severe health effects over time than that produced by physical violence (Mayhew *et al*, in press; Mayhew and Chappell, 2003; McCarthy *et al*, 2003b). This evidence is of crucial importance in any industry sector where there is already a high incidence of occupational violence as the cumulative burden can be debilitating.

In sum, because employment in advanced industrialized countries is increasingly skewed towards the services industries – which generally require extensive face-to-face contact between workers and clients/customers – an increase in the incidence and severity of occupational violence can be expected. In addition to the physical trauma following overt violence, the cumulative emotional/stress impact may take a heavy toll. The broader education sector (including tertiary education, school systems and juvenile justice) is not immune to these risks, particularly for workers who routinely deal with adolescents who are under stress for whatever reason.

Without widespread implementation of preventive interventions across the education sector, both the *incidence* and the *severity* of occupational violence/bullying experienced by those who work with adolescents will inevitably increase. There is a range of preventive interventions that are known to be effective in other industry sectors and these need to be adapted and adopted across the education system. Otherwise, the significant physical and emotional stress injuries will shortly reach epidemic levels.

9
Preventing Risk/Severity Spirals: A Cross-Sectoral Approach

Paul McCarthy

Incidents of workplace bullying/violence commonly manifest from an intensification of risks in interactions across individual, organizational and societal situations. The interplay of these risks can also contribute to the severity of impacts (as discussed in Chapter 8). Both risks and impacts can escalate over time for many forms of bullying and violence. This chapter focuses on the escalation of risks and impacts for incidents with lead in histories, repetition of incidents and post-incident interactions.

First, the relations of risk and severity for bullying/violence are examined as a basis for scoping the preventive challenge (Figures 9.1 and 9.2). Risk 'intensities' that can increase the likelihood of events and the severity of their impacts are then profiled and bullying and violence are identified as risk factors for each other. Case analysis of experiences of plaintiffs seeking legal remedies for workplace bullying/ violence informs the modelling of a 'typical' risk/severity spiral (Figures 9.3 and 9.4). Survey evidence of recipient's experiences in seeking help is also considered. Finally, a framework for the auditing of capacities of individuals, organizations and governmental interests to prevent risk/severity spirals is presented to further the development of a cross-sectoral web of safeguards (Tables 9.3a–f).

The analysis of preventive needs and strategies is schematic and applies a critical research methodology to case, practitioner, research, and advocacy experience (Hatcher and McCarthy, 2002). The notion of 'incivility spirals' (Andersson and Pearson, 1999) has provided the basis for modelling of risk/severity spirals for bullying/violence. Glasl's (1994) depiction of bullying as an escalated, destructive phase of conflict with potential for violence and self-harm has also informed the analysis.

Relations of risk and severity

The severity of an aggressive act can be analysed in terms of its *potential* to produce a physical or psychological impact. Alternatively, the severity of the impact of an incident of bullying/violence *actually experienced* by the recipient can be mediated by several factors. For example, the recipient's physical and psychological robustness, attributional processes and access to remedies, all have some bearing on the severity *experienced*. The extent of repetition of behaviours that are less overtly physical also has bearing on the severity of impacts (see: Chapter 8). For example, the experience of repeated bullying was found to produce traumatic impacts rivalling or exceeding those produced by rape and the encounter with violent suicides, for example by train drivers (Leymann and Gustafsson, 1996).

A review of case material indicates that a multiplicity of factors drive the escalation of risk for bullying/violence. The experience of bulling is often part of the incubation process for more overt violence. An experience of overt violence can also have its sequelae in bullying. Where risks intensify to the point that overt physical violence manifests, then the risk/severity spiral is likely to accelerate steeply (Figure 9.1). Whereas, in definitional terms, risks and severity of bullying incubate over time and then escalate and decline through similar degrees of post traumatic impact as that induced by overt violence. Observations of the interactive nature of risk and severity of workplace violence/bullying have lead to their treatment as mutually furthering phenomena in the analysis that follows.

The severity of impact of an overt physical attack is assessable in medical terms. The psychological severity of impacts of physical

Figure 9.1   **Severity of impacts of workplace bullying and violence over time**

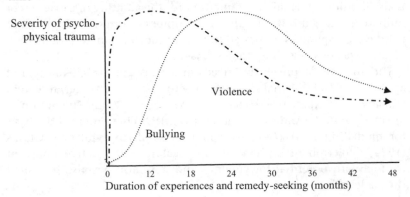

violence and threats of violence are measurable using criteria such as those for PTSD, General Anxiety Disorder and the General Health Questionnaire (Leymann, 1996). Apart from the initial impacts of an incident, trauma can escalate as the 'recipient' progresses to the status of 'victim' with the attribution of meaning to incidents. The traumatic impacts of repeated bullying and some forms of occupational violence accumulate over many months, sometimes years. Stresses in seeking remedies may also add to the trauma. Evidence also suggests that the longer bullying is experienced, the greater the severity of impact (Einarsen and Mikkelsen, 2003). Over time, the experience of bullying and PTSD effects may induce enduring personality changes, particularly in relation to depression and obsession (Leymann and Gustafsson, 1996). In the interests of providing a definitional cut-off for research purposes, Leymann (1997) required the recipient to experience negative actions at least once a week, for at least 6 months, to be classified as 'mobbed'.

Reduction of the duration of 'experiences' of bullying/violence is a key to breaking the risk/severity spiral. 'Experiences' need to be conceived of in terms of immediate impacts and post-traumatic impacts that accumulate after the incident. A study by Einarsen *et al* (1999) found that 75% of the victims of long term bullying experienced PTSD symptoms, and that the symptoms endured for more than 5 years after the bullying had ceased for 65%. Post-traumatic effects of bullying have been conceptualized in terms of symptoms *analogous* to those of PTSD, since most experiences of bullying do not entail an encounter with a life threatening event, as is required by the DSM-1V-TR (APA, 2000). This disjuncture has been met by a proposal that the traumatic effects of bullying be conceived of as *Prolonged Duress Stress Disorder* (PDSD) (Scott and Stradling, 1994).

A progressive disintegration of the conceptual framework through which the recipient may make sense of their life-world can occur over the time bullying/violence is experienced (Wilkie, 1998; Janoff-Bulman, 1992). Consequently, recipients may have diminishing capacity to cope with their experiences, work demands, remedy seeking, and extra-work social and family relationships. Niedl (1996) also found that victims are progressively less able to use problem-solving strategies. Social ostracism that may accompany bullying/violence has been linked with a range of negative effects, including: social belonging; self-esteem; desired image; sense of meaningful existence; and degrading of the immune system and health problems (Einarsen and Mikkelsen, 2002: 139–140). A recent study also found that victims of bullying regarded themselves as '... less worthy, less capable, and less caring' than others in a control group. Furthermore, the respondents perceived the world as '... less benevolent, other people as less supportive and caring, and the world as less controllable and just' (Einarsen and Mikkelsen, 2002: 138).

Risks of severe impacts are also borne by forms of bullying/violence at work that are systemic, epistemic and ethically-proper (Copjec, 1996). These types of violence are normalized in formations in language and culture and often perpetrated for the greater good. As such, they are subscribed to by many, wittingly and unwittingly. Processes of naming, labelling and excluding in epistemic violence routinely authorize behaviours and work practices that carry severe psychological and physical risks. Epistemic violence in the construction of the Other is also at the roots of gang warfare, communal violence, racism, genocide and terrorism (McCarthy, 1994). As Debray (1983: 297) observed, '... give me a good enemy and I will give you a good community'. Thus awareness and debate about the translation of meaning, concepts and labels into violence through conduits in thought, language, systems, emotions, organization and community are needed break the risk/ severity spiral.

The interaction of risk factors across individual, organizational and wider societal levels in the escalation of the likelihood of events and the severity of their impacts is profiled in Figure 9.2. This profile provides a basis for the depiction of 'intensities' through which risk and severity commonly escalate in the next section.

Risk intensities

Four risk 'intensities' considered prevalent in the escalation of risk and severity of bullying/violence in contemporary organizations are set out below. The notion of 'intensity' is used to illustrate the dynamics through which risk and intensity escalate, or spiral, due to the confluence of diverse forces.

Intensity 1: Intensification of risks through pressures of global market capitalism, restructuring, displacement, fear, and skills failures

Global market capitalism impacts on organizations through technological change, structural adjustment, economic rationalism, restructuring and outsourcing in ways that enhance the risks of bullying/ violence (Pusey, 1992; McCarthy *et al*, 1995). The relentless progression of modernism into culture precipitates the melting of 'all that is solid'

Figure 9.2 Intensification of risks and severity of impacts of bullying/violence at work

Individual risks	Organisational risks	Societal risks
Predisposition to aggression or provocation, poor impulse control	Normalisation of aggression	Economic restructuring; winding back of welfare; degraded social capital; human rights subsumed to market economy
High expectations of benevolence, justice; excessive conscientiousness	Competitive, high stress work culture; unreasonable work practices	Cultures of anxiety, fear, blaming, intolerance, and litigiousness
Negative attributional framework; negative affectivity; low self efficacy	Serial restructuring with poor people management	Limited regulatory framework, difficult access to tribunals, courts
Limited interpersonal, conflict, or anger management skills	Under-resourcing and equipment failures	Lack of transparency; distortion of information and intelligence
Limited coping and resilience capacities	Extremes of authoritarian or laissez faire management styles	Endemic corruption
Vulnerability due to personal or racial characteristics or life circumstances	Low tolerance of diversity	No recourse for breaches of professional, public sector or corporate ethics
Life or relationship stressors	Negligent environmental design	Stifling of public interest reporting; little protection of whistleblowers
Precarious employment	'Hot' products, isolated, unprotected worksite, night work	Politics of fear; sacrifice of the less-powerful for the greater good
Hypervigilance and revenge seeking due to previous difficulties	Lack of zero tolerance policy	Religious fanaticism; celebration of the victim/martyr
Poor mental health; alcohol, substance abuse	Limited complaint, reporting, developmental and/or disciplinary mechanisms	Lack of governmental obliging and rewarding of prevention
Negative, unhealthy identity	No independent conflict resolution	No alignment of preventive regulations to enhancement of competitiveness
Neuroses; paranoia; psychoses	Poor data/reporting system	
Callous un-emotionality; masochism; sadism	No accountability for impacts on productivity, quality, customers or investors	
	Costs displaced to insurers	
	No CSR reporting	
	No vulnerability auditing	

Plus   *Plus*

⇨

Intensification of risks

and

Higher likelihood of incidents of bullying/violence

and

Escalating severity of impacts

(Berman, 1982) with consequent loss of cultural, social, and conceptual grounding. Dislocation of communities and work and family futures contributes to a generalized climate of fear and anxiety. In this scenario, pressures at work, fears of job loss and consequent threats to status and consumption, all resonate with personal anxieties about ill health, violent crime, environmental degradation, and relationship difficulties.

Work technologies and the labour process are subjected to continuous change. Cost cutting is widespread and overworking is the norm amongst core employees, while many are precariously employed. Many managers and workers are pushed to the limits of their skill competencies in these circumstances (Karpin, 1995). Intensification of work and life stresses lower the threshold at which managers, employees and clients can reflex into aggressive behaviours (McCarthy *et al*, 1995). The experience of bullying is also a risk factor for more overt violence in this scenario, as has been demonstrated in the 'going postal' homicides and massacres. The Oklahoma bombing stemmed from the potent infusion of economic disadvantage, political righteousness and religious fanaticism (e.g. 9–11). Thus, experiences of bullying and disadvantage can be a trigger for terrorism (Barker, 2003; McCarthy, 2003a).

Intensity 2: Intensification of risks through new formations of meaning, attribution, injury, evidence, and litigiousness

The attribution of meanings of bullying/violence to the experience of negative actions and distress at work provide the basis for the taking of 'victim identity' and the pursuit of remedial pathways (Lewis, 2000; Liefooghe and Olfasson, 1999; McCarthy, 1999). In the taking of victim identity, risks of bullying/violence can intensify because of difficulties in defining, reporting, evidencing, and accessing remedies. This is particularly the case where hitherto normalized behaviours are challenged as bullying/violence as awareness increases with the promulgation of guidelines or regulations.

The identity, behavioural characteristics and psychosomatic health of the 'victim' of workplace bullying/violence is socially modelled and legitimated by self-help books, victim-support groups, counsellors, equity and harassment officers, union officials, and lawyers. Meanings in guidelines and regulations for the prevention of bullying/violence also shape the character of the victim. Fluidity in these meanings due to perceptual, definitional and evidencing difficulties may contribute to stresses in pursuing remedies in grievance proceedings, tribunals and civil courts. These stresses may act to compound the severity of impacts, and give rise to risks of further experiences or reactions of a bullying or violent nature (Hatcher and McCarthy, 2002).

Intensity 3: Risk intensification through personal and organizational vulnerability, distress, coping, complaining and blaming

Vulnerabilities and lack of coping capacities at the individual and organizational levels are the weak links from which risks intensify. Unreasonable work systems/practices heighten the likelihood of bullying/violence. Prior socialization in dealing with bullying/violence and experiences of difficulties in having complaints addressed may inhibit recipients from complaining about incidents. Similarly, recipients often fear that speaking out will jeopardize their job futures, prompt payback, and can be daunted by the cost and emotional energy that has to be invested in remedy-seeking.

Socio-culturally specific ways of coping with bullying/violence may also predispose recipients to internalize, complain, or pass on the distress of bullying/violence. Where there are perceptual, definitional and evidencing difficulties, recipients who do speak out also risk becoming caught in a vicious 'blame game' that compounds the risks, costs and severity of the impacts of bullying/violence. There is also the risk that, under pressure, an employee, customer, sub-contractor or bystander who has a history of a 'short fuse', may experience vulnerability, even neuroticism, paranoia or psychosis that flares into spiteful or violent counter-reactions.

Case evidence indicates that bullying/violence leads recipients to take days off work and causes their performance and relationships to deteriorate. The psychosomatic sequelae can be so debilitating that the victim may remain absent from work without gaining the required medical certificate. Unauthorized absences, problems with work output, and difficulties with other staff members can be magnified into charges of unsatisfactory performance and misconduct. With the advent of policies and procedures to prevent bullying/violence, allegations of unsatisfactory performance can be used (instead of bullying) to weaken the victim's position. These responses conspire to hide the behaviour of the perpetrator as manager within the domain of 'reasonable management practice' (e.g. DIR Qld, 2004), hence not a matter for consideration within the anti-bullying/violence policy. Investigators and mediators can also be from within the social and professional orbit of the perpetrator/manager and bias the mediation process against the recipient.

Medicos commonly attribute the distress of all but the most overt forms of bullying/violence at work to individual pathologies and prior psychological histories, effectively 'blaming the victim' (Lennane, 1996). Although OSH legislation usually requires incidents at work occasioning injury be reported, and can obligate medicos to report incidents they treat, most incidents of bullying/violence are not formally reported (Mayhew, 2002a; Chappell and Di Martino, 2000). However, some incidents are reported when a workers' compensation claim is made, although agencies do not often publish aggregate claims data. Thus, preventive strategists are deprived of a reliable source of information in which to base strategies.

Claims for workers' compensation due to overt physical violence that produce visible injuries or PTSD are more likely to be accepted by insurers, where medicos can confirm clear diagnoses. However, psychological injuries consequent to bullying are often less able to be witnessed or verified as predominantly caused by the bullying, since the offending behaviours are more subtle and can be systemic. Persons who make claims for workers' compensation due to bullying/violence at work are often subjected to long periods without income while claims are processed, verified, accepted/rejected and subjected to appeals. This loss of income can undermine capacities to mount legal challenges against the rejection of worker's compensation claims and common law actions in industrial or civil courts. During these periods, a complainant's psychiatric profile can worsen dramatically, symptoms of PTSD can take hold, and risks of experiencing new incidents of bullying/violence and aggressive counter-reactions ensue.

Evidence suggests that the majority of persons who pursue legal compensation for bullying are unhappy with their experience (Barker and Sheehan, 2000). 'No-win, no-fee' financing of actions by lawyers also contains risks that settlements or judgements in favour of the plaintiff will be consumed by legal fees. In cases that plaintiffs have lost 'no win, no-fee' financed legal claims for damages induced by bullying/violence, or won judgements or settlements smaller than actual legal costs, lawyers may still hold plaintiffs liable for costs.

Intensity 4: Risk intensification through victimization, revenge seeking, border protection, and homeland security

Feelings of being 'under siege' pervade contemporary 'risk society' and victim-mentality can take hold (Bauman, 2002; McCarthy, 1999; Beck 1992). Threats invoked by terrorism and illegal immigration resonate with fear of displacement and job loss in organizations restructuring to meet global market exigencies. A common consequence of this welling of fear is in the projective construction of the Other as the source of all ills, and thus deserving of bullying/violence. This projection of fears catalyses the hardening of emotional arteries, identities, team and organizational boundaries, and borders. The risks of bullying/ violence evoked by these sensibilities intensifies along tangents of fear, the hardening of identities and borders, revenge, and subjugation (Chomsky, 2001). Generalized fears intensify with angst evoked by encounters with callous co-workers, managers, sub-contractors, suppliers and customers. This angst can be projected into disaffection with organizations, civil society and the political system, and fuel neo-conservatism driven by racist anti-immigration sentiment.

In this climate, intersections between victim-mentality, personal vulnerability, and political and religious righteousness unleash risks of bullying/violence against the enemy within and without (McCarthy, 2003a). The experience of bullying/violence in this climate tends to be followed by post-traumatic stress coupled with an obsessive litigious pursuit of reparation and protection. The strategic labelling and surveillance of the Other as a threat can be legitimated to the point that basic human rights and liberties are foregone.

Escalation of risk and severity in seeking remedies

An analysis of cases of bullying/violence involving legal action has been completed to provide a basis for the development of safeguarding strategies that prevent the escalation of risk and severity (see: Table 9.1). The sources of the cases are confidential and aggregate features of each case have been presented in a manner that does not enable them to be identified. Doing nothing about experiences of bullying/violence, or trying to 'tough out' situations may of course leave recipients exposed to further incidents and increasing trauma. However, as the analysis in Table 9.1 suggests, the added stress of seeking remedies may act to compound risks of further events and severity of impacts.

The duration of experiences in the cases profiled in Table 9.1 averaged 41 months from the first experience of bullying/violence until the approximate date of legal settlement. The number of persons involved in each case varied from 14 to 39 (average 23). More people became involved as the victims progressed up the remedial hierarchy. Factors that compounded the severity of impacts included concurrent experiences of: unreasonable behaviours and work systems; client abuse; ongoing interactions with perpetrators; fears and actual experiences of further incidents; accumulating costs; and the growing burden of inter-actions in seeking remedies. Of course, the distress continued well beyond the cessation of the legal actions.

In the cases reviewed in Table 9.1, interactions with perpetrators (all fellow employees), managers and supervisors, witnesses, workers' compensation agencies, assistance providers, medicos and lawyers were ongoing while matters remained unresolved. These interactions involved continual revisiting of traumatic incidents as claims were re-stated to a growing number of actors. More than 4 doctors (general practitioners or psychiatrists) became involved in each case on average. Victims reacted to the perpetrator inappropriately in two of the cases, as frustration and trauma escalated, and were themselves subjected to complaints of bullying/violence (as a consequence, both these victims were found unsatisfactory in performance appraisal). In four of the 6 cases, the victim's family members expressed concern and anger to management about the impacts of the bullying/violence. There was little significant interaction with union representatives, and in Case 6, the victim felt inhibited because the union representatives were aligned with the perpetrators; and interactions with counsellors were uncommon.

Table 9.1 Interactions in cases of bullying/violence involving legal action

Case Features	Case 1	Case 2	Case 3	Case 4	Case 5	Case 6	Case average
Duration (Months)	25	48	52	23	42	56	41
Experiences*	B, S	B, S	B, C, S,	B, S	B, V, S	B,V,S,C	
Perpetrators**	2	2	3	1	1	6	2.5
Witnesses**	3	6	5	4	4	7	5
Managers & Supervisors**	4	8	1	3	2	6	4
Family**	3	1	2	1	3	3	2.2
Union reps**	0	1	0	0	0	2	.5
Consultants **	1	0	2	0	0	3	1
Counsellors/ EAP**	1	0	0	0	0	1	.3
Medical doctors**	6	4	6	3	3	6	4.7
Worker's comp***	1	1	1	1	1	2	1.2
Law firms***	2	2	2	2	2	4	2.3
Total actors	23	25	22	15	15	39	23

* B: bullying; V: physical contact or threat; C: abuse from clients; S: unreasonable work systems and practices
** Numbers of persons according to row category
*** Number of agencies or organizations according to row category

Where the bullying/violence occurred in smaller privately owned organizations victims endured fewer interactions (e.g. Cases 4 and 5, Table 9.1). These organizations featured smaller numbers of staff, less bureaucracy, and fewer personnel, equity or harassment officers empowered to intervene in issues of bullying/violence. They also lacked policies that provided a vocabulary for naming experiences and seeking remedies. In contrast, recipients in larger organizations bore the burden of increased interactions while pursuing remedies. In one case, both the recipient and the alleged perpetrator left the employer believing the other was the perpetrator, and each later initiated legal action against the employer.

Implications from the review of cases in Table 9.1 are that preventive mechanisms need to alleviate the growing weight of interactions and added stresses that manifest in the pursuit of remedies. Empowerment of those in the first line of supervision to resolve conflicts when and where they first happen is the most immediate way to head off the escalation of risk/severity. Where concerns cannot be promptly resolved, recipients/victims need to be supported and represented as they progress up the hierarchy of remedies so as to minimize frustration, duration, cost, added stress, and escalating severity of impacts.

Risk/severity spirals

The model risk/severity spiral presented in Figure 9.3 compresses the interplay of risk and severity factors for bullying/violence into six 'typical' stages of escalation. The possible duration of these stages from the first experience of incidents is also indicated. The stages are not mutually exclusive. Progression through these stages by a recipient may be delayed, reversed, or accelerated according to mediating factors in the individual, organizational and societal context. Remission may ensue where complaints are resolved and the victim progresses to 'survivor'. Alternatively, there may be a doubling of risks and severity when new incidents manifest and complaints, grievances and legal claims are not resolved.

Stage 1: Exposure to negative experiences that bring vulnerability factors into play for the individual and the organization (Up to 6 months)

Work systems and practices that over-stress, overload, under-resource and fail to protect employees increase risks of exposure to bullying/violence. Individuals who carry failures to cope with prior experiences of stress, conflict and depression into a new work situation may be

Figure 9.3   **Model risk/severity spiral**

more susceptible to negative impacts of unreasonable behaviours and work systems. Negative affectivity, low self-efficacy, over or under conscientiousness, and unreasonable expectations of benevolence and justice have all been recognized as individual risk factors for bullying/violence (Einarsen, 2000a; Zapf and Einarsen, 2003). Employees with aggressive predispositions may well contribute to conflicts that are precursors to bullying/violence. Authoritarian management styles can also be a source of conflict. On the other hand, *laissez-faire* styles may leave conflicts between competing work teams and individuals unresolved (Hoel and Salin, 2003). Workplaces that feature poor environmental design are vulnerable to 'external' instrumental violence (see: Chapter 2) (Caple, 2000; Jeffery & Zahm, 1993; Clarke, 1992).

Stage 2: Failure to cope, naming the experience as a form of bullying/violence, and taking the position of victim

Recipients may be inclined to name the events, attribute blame, and identify perpetrators according to their awareness of bullying/violence and prior experiences with complaints. Organizations with policies that explicitly address forms of workplace aggression also provide employees with literacy to name negative actions as 'bullying' or 'violence'. The extent to which the recipient adopts the position of victim can also reflect their exposure to social modelling of victimhood, in other victims, preventive policies, and therapeutic discourses. The organization's response to complainants also shapes the victim's identity. Recipients who confront perpetrators or make complaints face risks of escalation of conflict and labelling as troublemakers. Internal complaints mechanisms may also inculcate bias since investigators and mediators are often in power relations that work against finding fault with the organization or senior level perpetrators. Thus there are many reasons why recipients who take the position of victim can become 'stuck' in unhealthy identity and symptomatology prone to escalation of impacts.

Stage 3: Complaint failure, payback for complaining, absenteeism and workers' compensation claim

Where prompt resolution of complaints or grievances to the satisfaction of all parties is not achieved, the recipient may be exposed to further bullying, stigmatization, and unsatisfactory performance appraisal. The extent of union support in complaints/grievances is contingent upon union membership. Rising precarious employment mean that fewer employees have organized support in seeking remedies.

Access to union support in complaints/grievance procedures can be difficult where union representative is a perpetrator or aligned to perpetrator/s. During this stage, the victim's anxiety and depression commonly intensify to the point that they take days off work. Medical officers and psychiatrists are commonly consulted for treatment and proof of injury for absenteeism and workers' compensation claims. However, diagnoses can be distressing where the cause of the bullying/violence is attributed to the victim's personality, and the perpetrator's behaviour or unsafe work systems/practices are ignored. Anti-depressants and mood enhancing drugs are often prescribed to alleviate symptoms (McCarthy and Rylance, 2001). Workers' compensation claims for psychological injury induced by less overt forms of bullying/violence are common, but difficult to prove and prone to rejection on the basis that the individual's prior pathology was responsible. This tendency to individualization of broader organizational and societal causal mechanisms has been widely noted for other OSH conditions, notably during the RSI 'epidemics' of the 1980s.

Stage 4: Difficulties in working, pursuing legal claims, job loss

Where PTSD symptoms due to bullying/violence at work take hold (Leymann, 1996) many are too traumatized to maintain their position at work. Where victims do take time out of work on compensation to deal with psychological injuries, the quality of their rehabilitation and return to work programs have considerable bearing on their job futures (Kearns *et al*, 1997). At this stage, the victim's decline in job performance or retribution can also lead to them being forced out of their job due to unsatisfactory performance appraisal. The threat and actualization of job loss during this stage commonly motivates inquiries about options for legal action. Information about costs, evidential needs, and duration is commonly sought for appeals against the dismissal of workers' compensation claims, applications in human rights or anti-discrimination commissions, and claims in industrial courts for breaches of the employment contract and unfair dismissal.

While legal remedies for overt forms of violence may be accessible via the criminal courts, difficulties remain in proving less overt negative actions by perpetrators, or unsafe work practices, were the predominant cause of psychological injury. The accessing of legal remedies is time consuming and involves the painstaking recall, documenting and witnessing of incidents and their attributed causes and effects. The added trauma and cost entailed also acts to sustain and compound the severity of impacts.

Stage 5: Symptoms of severe psycho-physical injury and relationships breakdown

Many victims with symptoms of PTSD find it difficult to persist in paid employment (see: Table 9.1). Experiences of financial hardship due to loss of income and mounting medical and legal costs add to the trauma. At this stage, victims are also prone to suffer breakdowns in their relationships with partners, friends, and work mates. Ongoing encounters with victims of bullying/violence can be exhausting and alienating to all but trained therapists, due to the victim's overwhelming misery, depression, fixation with retelling of their story, alternation between anxiety, depression and anger, hyper-vigilance, and suicidal state (Einarsen and Mikkelsen, 2003). The consequent loss of social grounding also compounds the victim's trauma.

Stage 6: Permanent displacement from work, obsessive seeking of justice, revenge, suicide

Where victims are damaged to the point that their capacity to engage in paid work and normal social interactions is permanently impaired, they suffer loss of income and increased costs of medication and psychiatry for indefinite periods. In financial terms the costs can be substantial, and the victim's plight may attract the attention of legal firms that pursue claims for damages due to negligence and breach of the employment contract in civil courts. Claims for damages in excess of £.3 million are not uncommon. In some cases, 'no-win no fee' arrangements are entered into. Often, expert witnesses in these cases give opinions that the victim will never be capable of resuming their pre-injury careers. These action for damages can endure for years and entail appeals. In some cases victims are so damaged that they are unable to testify.

The stress of displacement from paid work and damages claims, plus disappointments with judgements, and bills for legal costs that substantially reduce or exceed awards of damages awards or settlements, all compound the trauma. Where victims are predisposed to revenge seeking, or network with those that are, there are heightened risks of violent attacks on perpetrators or property of the organization. Extreme anger and revenge seeking can also alternate with suicidal ideation where hopelessness pervades (Leymann and Gustafsson, 1996).

The six stages of risk/severity escalation for a broad range of experiences of bullying/violence that originate with the progressive failure to resolve workplace conflicts are depicted in Figure 9.3. Random violence from external sources (e.g. hold-ups) does not immediately fit within the model. Although, where the organizational footprint degrades community life and employment to the point that ready cash for consumption and drugs assumes over-riding importance, all organizations in a locality or region can be at risk. In Figure 9.3, entry to the risk severity spiral is at the bottom left hand corner. The extreme end point in the risk/severity spiral in Figure 9.3 is one of elevated risk for suicide, homicide, 'going postal' with mass homicide, or terrorism.

The generalized model of the risk/severity spiral presented in Figure 9.3 is applied schematically in the case analysis in Figure 9.4. The specific case events depicted in Figure 9.4 are a pastiche of 'typical' events that mark the progressive escalation of risk and severity in a 'worse case' scenario of bullying/violence.

Help seeking

The risk/severity/spirals outlined in Figures 9.3 and 9.4 indicate that access to remedies for bullying/violence are difficult and it is not surprising that most incidents also go unreported. In reality, most recipients deal with bullying/violence through personal coping and may pursue help from a variety of personal and professional sources outside the organization. Constructive experiences in gaining help have the potential to reverse the risk/severity spiral.

Many organizations lack accessible, workable policies and procedures, and in many situations complaining is realistically assessed as too risky. In these circumstances, help from personal/social sources and service providers may be the only means of preventing the slide from 'recipient' (i.e. alert but not distressed) to unhealthy victim-identity (distressed, in flight/fight mode, and panicking). The quality of help from personal and social sources appears to be proportional to the provider's understanding of the issue. Help from market-based service providers reflects disciplinary approaches that can be more or less helpful to victims. For example, inciting a recipient to make a complaint in risky circumstances that leads to job loss may not be what they really 'need'. Nor might the recipient/victim really 'need' to take on long drawn out grievance or legal proceedings of indeterminate cost and duration for less common or severe events, given risks that the added duress from the proceedings could compound the trauma.

In all cases there is a danger that, in seeking help, the 'recipient' can be incited to take on dimensions of victim-hood that reflect

Figure 9.4   'Typical' case events in the spiralling of risk and severity

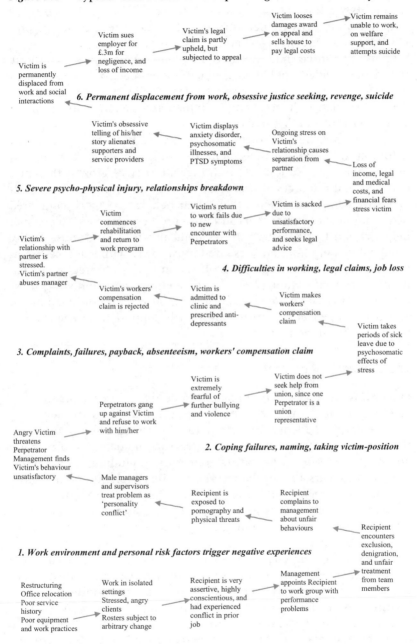

resentments and world views that are less than healthy. Help for recipients/victims of bullying and violence is more likely to be constructive and healthy when informed by research evidence, is provided by independent professionals, and is accessible.

Results of a survey into sources of help outlined in Table 9.2 indicate that many recipients of bullying/violence are proactive in seeking seek help from a wide variety of sources (McCarthy and Zimmermann, 2002). Experiences in accessing help mediate the recipient's progression to 'victim' and to 'survivor'.

The data in Table 9.2 shows that recipients of bullying typically seek help from workmates, management, the union, medicos, lawyers, and psychologists. Education and training, alternative growth therapies and personal resources also provide sources of help. At work, workmates were most often sought and their help rated highly, compared with poor ratings of managers and union representatives. Professional help was most often sought from medical officers and most highly rated, while help from lawyers rated most lowly. Other sources nominated as helpful (by proportions of respondents indicated) included: training and education (77%) with 'reading about bullying' most common; alternative therapies (34%) with meditation most common; and personal/social resources (90%) with support from friends and family most common.

The challenge of prevention

Preventing the escalation of risk and severity as risk factors interact across individual, organizational and societal levels is at the core of the preventive challenge. Here 'individual' encompasses not only employees of an organization and its directors, managers and owners, but others within its operating footprint, including clients, contractors, suppliers and regulators who can be recipients and initiators of bullying/violence in interactions with the corporation. The 'societal' context is also short hand for the socio-economic, civil, legal, political and governmental assemblages within which the organization operates.

In preventing the escalation of risk and severity, we need to be mindful that perpetrators do not always stand out as sadists, psychopaths or sociopaths. Perpetrators may be the 'normal' people identified by Bauman (1991) who adopt roles through which bullying/violence are transmitted to the utilitarian benefit of the greatest number. Experiments by Milgram (1965) and Zimbardo (1988) demon-

Table 9.2 Experiences of seeking help by victims of bullying (n = 53)

Sources of help	Proportion seeking help	Extent of help gained
Workplace		*% finding source most helpful*
• Workmate	71%	33%
• Union	57%	12%
• Managers	51%	8%
Professional		*% finding source most helpful*
• Medical officer	63%	44%
• Legal practitioner	53%	27%
• Psychiatrist	36%	33%
• Psychologist	35%	36%
Training and education		
• Read about bullying	85%	*77% of respondents seeking*
• Assertive training	39%	*help through training and*
• Communication skills	34%	*education indicated found it*
• Conflict management	31%	*helpful*
Alternative therapies		
• Meditation	50%	*34% of respondents seeking*
• Acupuncture	33%	*help from alternative growth*
• Massage	33%	*therapies indicated found*
• Psychic	22%	*them helpful*
• Yoga, Tai Chi, Herbalist	16%	
Personal/social resources		
• friends support	77%	*90% of respondents seeking*
• family	70%	*help from personal and social*
• books & reading	60%	*resources indicated found*
• pets	34%	*them helpful*
• leisure	27%	
• sport	25%	
Other sources		
• writing & researching	16%	*85% of respondents seeking*
• reported, legal advice	13%	*help from other sources*
• medication, substances	11%	*indicated found them helpful*
• confronted bully	11%	
• prayer	11%	
• strength, determination	11%	

strated the potential for 'normal' citizens to relay cruelty. That is, family loving ordinariness is no protection against the transmission of brutality – as has been observed among personnel in the Nazi apparatus, and amongst professional torturers (Bauman, 1991; Conroy, 2000).

There is also the phenomenon of the 'bully-victim' in which positions of victim and perpetrator crossover. Arguably, this crossover is pervasive in postmodern organizations and society (McCarthy, 2000a,b). The labour process in new flatter, flexible organizations that continually transmogrify to meet global market conditions invariably produces bullying/violence (Casilli, 2000). The serial restructuring and cost cutting in the new organizational forms inevitably relay all sorts of little brutalities that marginalize, threaten, and exclude. Pressures lead the new work teams to function as 'small furnaces' fuelled by micro-fascisms (Guattari, 1984). There are no pure positions, and we are all more or less complicit as 'normal' individuals work to further personal survival and lifestyles geared to consumption of symbolic goods (Hatcher and McCarthy, 2002). Hence, bullying/violence may become a normal part of interactions.

Recipients of bullying/violence commonly neither report incidents, nor have complaints dealt with to their satisfaction. Lack of evidence, perceptual differences, fears of retribution or stigmatization, and/or lack of policies/procedures, are all reasons that reduce reporting. Nevertheless, workplace bullying/violence poses an enduring threat to the civility, safety and trust necessary for the favourable conduct of business, government, and community life in social democracies. While overt violence and threats of violence are criminal offences, that does not ensure that many forms of occupational violence will be reported or sanctioned (Mayhew and Chappell, 2002). The many forms of less overt workplace bullying/violence are less easily regulated.

The prevention of risk/severity spirals is not as simple as identifying discrete acts, prohibiting them at law, and punishing perpetrators. Given 'bullying at work' has arisen as a new syndrome of distress that is perceptual and difficult to define and evidence, there is a danger of crusading, witch-hunting, and punitive legalism that is misplaced (McCarthy, 2003a). We argue that prevention of bullying/violence at work is a process of long-term cultural change. A cross-sectoral approach to progressive development of awareness and commitment to regulations and obligations geared to restorative justice and recon-ciliation (Ahmed *et al*, 2001) over a number of decades are necessary to achieve this change.

The preventive challenge expressed in Figure 9.2 reflects the reality that individuals, (e.g. employees, clients, contractors, and community members) need to fully understand the operation of preventive policies within organizations and the wider regulatory framework. It is also the case that organizations are unlikely to implement preventive policies and procedures where not obliged by regulatory compliance standards (McCarthy and Barker, 2000).

Definitions of bullying and other forms of occupational violence can be inscribed in OSH industrial legislation. However, difficulties in defining and evidencing bullying and some less overt forms of occupational violence means that, while necessary, legislation against bullying is not likely to be a sufficient prevention on its own. Governments can best fulfil their obligations to ensure the quality of both social and productive life by providing access to external preventive measures that also oblige responsible behaviour by individuals and organizations. However, such remedies are unlikely to gain the political purchase that ensures commitment by government necessary for their effective operation, unless they garner widespread industry, union, professional and community support.

The circular weakness in preventive logic depicted so far brings us to the contention that effective prevention is contingent on cross-sectoral commitment to obligations to prevent bullying/violence that is motivated by realization of mutual benefits (Zadek *et al*, 2003). The governmental and community endorsement of prevention requires recognition of advantages in terms of safety, security, productivity, fair trading, corruption-prevention, social responsibility, accountability and hence, overall competitiveness. To ensure the realization of mutual benefits, the regulatory framework needs to promote collective subscription to measures that prevent 'rogue' employers using bullying/violence to extract a short-term market advantage.

Apart from reducing risks of events in the first instance, the challenge is to prevent the compounding of trauma due to fears of the repetition of incidents and added distress in seeking remedies. There is also the risk that bullying can trigger violence. Clearly, allegations of bullying/violence need to be resolved promptly and fairly to avoid the escalation of risk and severity over time. In these terms, the challenge is to prevent the intensification of risks to the point that:

- Intermittent or random negative actions are repeated;
- Recipients and alleged perpetrators are denied natural justice in complaints resolution processes;
- Traumatic impacts following events are internalized by recipients to the point they become victims;
- Complaints/grievances remain unresolved;
- Legal actions have indeterminate cost, duration and outcome;
- Risks of bullying/violence escalate as recipients are unable to gain justice, perpetrator behaviours continue and revenge is sought; and,

- Victims enter a downwards spiral that ends in permanent displacement from work and suicide (Leymann, 1996).

A cross-sectoral web of safeguards

The stages in the escalation of risk and severity depicted in Figures 9.3 and 9.4 and the scoping of preventive needs in Figure 9.5 provide a basis for the *architecture of safeguarding* proposed in this section. The framework outlined addresses diverse forms and directions of bullying and violence through mutually supportive interventions at the individual, organizational and wider societal levels. The framework proposes an integrated approach to common risk factors for bullying/violence that incubate over time. The 'ideal' interventions considered necessary to arrest the escalation of risk and severity through each of the six stages identified in Figures 9.3 and 9.4 are:

- *Stage 1:* Assessment of personal, organizational, and societal vulnerability to inform risk management
- *Stage 2:* Promotion of constructive coping, zero tolerance policies and regulatory obligations
- *Stage 3:* Resolution of informal/formal complaints/grievances in ways that reduce absenteeism and workers' compensation
- *Stage 4:* Constraining the escalation of risk/severity to the point legal claims are launched and the victim is forced out of the workplace
- *Stage 5:* Managing the transition from workers' compensation to rehabilitation and return to work with positive relationships and ongoing attention to PTSD symptoms
- *Stage 6*: Moving on from legal action, suicidal ideation and revenge seeking, to learning, re-affirmation, reconciliation and constructive re-engagement with work and life

The checklists for the *Audit of Capacities to Safeguard* at each of the six critical stages of escalation of risk/severity (see: Figures 9.3 and 9.4) are presented in Tables 9.3a–f. This audit framework has been designed to prompt interventions at the individual, organizational and governmental levels that collectively optimize prevention while generating mutual benefits from enhanced productivity, health, security, and competitiveness. The selection and patterning of the interventions reflects the evolution of the author's thinking about bullying and violence in the past decade (see: McCarthy, 1993–2004

Figure 9.5   **Mutually supportive individual, organizational and governmental strategies to prevent bullying/violence risk/severity spirals**

Organisational
policies and procedures that provide:

- *Risk assessment and management*
- *Best practice environmental design*
- *Appropriate training, information, and confidential independent advice*
- *Commitment of employees, clients and contractors to codes of conduct and responsible behaviours*
- *Access to a hierarchy of remedies, including: informal and formal complaint resolution, grievance procedures, and opportunities for independent mediation*
- *Initial no-blame conflict resolution and restorative justice, with developmental remedies for both recipients and perpetrators and escalating penalties for serial offenders*
- *Support and protection for recipients/victims in complaints processes*
- *Linking internal policies/procedures with emergency, security, and intelligence services*

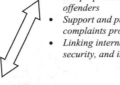

Individual
employees, clients, suppliers and contractors and stakeholders who are:

- *Aware of meanings and remedies for workplace bullying/violence*
- *Committed to organisational policy and procedures and the regulatory framework*
- *Responsible in assessing whether their vulnerability, lifestyles or behaviours contribute to risks of bullying/violence*
- *Motivated to develop their coping and resilience capacities*
- *Able to assess risks of taking the position of 'victim', proceeding with complaints, grievances or legal action, possibilities of retribution, and viability of remedies*
- *Engaged in constructive dialogues with management, employee associations, regulators, institutional and third-sector interests in the assessment of risks and implementation of preventive strategies*

Governmental partnering
with industry associations, unions, institutions, agencies and third-sector interests in:

- *Development of regulations that oblige the implementation of preventive policies*
- *Provision of external remedies through alternative dispute resolution, tribunals, and the ombudsman*
- *Obligation of local independent no-blame conflict resolution within workplaces before accessing external remedies*
- *Arrangements with third-sector organisations to provide information, support, counselling and dispute - resolution*
- *Recognition and rewarding of interventions (eg. in workers' compensation, tendering and licensing requirements)*
- *Promoting competitive and social advantages of prevention, eg. the contribution to productivity, safety and security, sustainability, corruption prevention, innovation, employment and community engagement*

and McCarthy *et al*, 1995–2003). The interventions also reflect indicators of good preventive practice in the literature (e.g. Di Martino *et al*, 2003; Einarsen *et al*, 2003; ILO, 2003a; Axelby, 2000; Mayhew 2000a,b,c).

Table 9.3a Audit of capacity to assess personal, organizational, and societal vulnerability to inform risk management

Individual capacity	Organizational capacity	Governmental capacity
Assessment of vulnerabilities ☐ Do my work locations, products, clients, co-workers, managers, or external parties place me at risk of bullying/violence? (E.g.: isolated locations, poor visibility and lighting; exchanges involving cash, 'hot' products, alcohol and drug abuse, physical/mental health problems, or financial and relationship difficulties) ☐ Does my work involve an unmanageable workload, impossible deadlines, or poor equipment and training? ☐ Do the personalities, skills or life circumstances of staff, contractors, and clients pose risks for bullying/violence? ☐ How developed are my capacities to cope with bullying/violence? (E.g.: understanding of risks, preventions and sources of help; and skills in communication, anger management, conflict resolution, stress management, self-defence, and first-aid) ☐ Am I predisposed to risky beliefs, moods, or behaviours?	*Assessment of risk management strategy* ☐ Are the CEO, management and staff committed to a zero-tolerance policy? ☐ Does scientific research inform risk assessment? ☐ Is top-down training in the policy/procedures routine? ☐ Is the policy aligned to international standards (e.g. ILO code, Human Rights, WHO)? ☐ Is safe environmental design applied to buildings, processes, and products to reduce risks (e.g. target hardening layouts, visibility, access, emergency response, and evacuation)? ☐ Are appropriate resources, training and work practices provided so that staff and clients are not aggravated? ☐ Is the anti-bullying/violence policy integrated into the policy framework? (E.g.: OSH; equity and quality enhancement; code of conduct; staff recruitment, selection, appraisal, performance review, development, and discipline; mental health and substance abuse; social/environmental reporting; and emergency response)	*Commitment to a cross-sectoral response* ☐ Do sectoral leaders at local, regional, state and national levels endorse zero-tolerance policies? ☐ Is consultation with diverse stakeholders used in developing the regulatory response? ☐ Do research partnerships investigate risks, prevalence, costs, and remedies? ☐ Are prevalent industry, societal, cultural, political, religious, and global predispositions to bullying/violence factored into the response? *Regulations and standards that oblige prevention* ☐ Do OSH, industrial, trade practices, governance, and consumer regulations and guidelines consistently define and mandate prevention? ☐ Does the regulatory framework comply with international standards (e.g. ILO, Human Rights, WHO)? ☐ Are built environment and transport design principles consistent with indicators for safety and security?

Table 9.3a **Audit of capacity to assess personal, organizational, and societal vulnerability to inform risk management – continued**

Individual capacity	Organizational capacity	Governmental capacity
(E.g. aggression, provocation, depression, low self efficacy, unrealistic expectations of benevolence/justice, excessive conscientiousness)	☐ Is the emergency response aligned to fire, rescue, ambulance, medical, and police responses, as well as intelligence and security agency surveillance and intervention strategies?	(E.g.: pedestrian access; public transport; secure people-friendly public space; shopping and entertainment areas; safe communities/towns programs; police beats; electronic surveillance; weapons detection; community observers and neighbourhood watch; and emergency response)
Assessment of risks	☐ Are the policy/procedures subject to ongoing monitoring and scientific evaluation?	☐ Is the reporting of preventive performance obliged within organizations and their operating footprint?
☐ To which patterns of bullying/ violence at work am I most vulnerable?	☐ Have regulations been mainstreamed in management practice to realize productive and competitive benefits from safeguarding?	
☐ In which areas are my response capacities and those in the workplace most deficient?	☐ Is the organization committed to conjoint industry and government sanctions against organizations using bullying/violence for short-term advantage?	*Rewarding prevention*
Personal safeguarding plan		☐ Are benefits accorded to contractors, customers and suppliers who implement preventive measures? (E.g. through favourable worker's compensation premiums, qualification for government tenders, and licensing requirements)
☐ Which work risks warrant reporting? (E.g. to safety officer, union, HRM, or external agency)		
☐ What activities could improve my capacity to respond to salient risks? (E.g.: personal or staff development)		

Table 9.3b **Audit of capacity to promote constructive coping, zero tolerance policies and regulatory obligations**

Individual capacity	Organizational capacity	Governmental capacity
Establish personal response options	*Information provision*	*Initiatives by cross-sectoral partners*
☐ Is the behaviour bullying? (E.g. repeated negative actions or unreasonable work practices, or violence, physical harm and/or threats)	☐ Is information provided via newsletters, booklets, posters, and web-site?	☐ Are information kits, web-site and personal advice available to employees unable to resolve workplace difficulties? (Particularly to non-union members and small-business employees)
☐ What is the likely severity of personal impact?	☐ Is bullying/violence differentiated from reasonable management practice?	☐ Do industry bodies, unions, and government and third-sector agencies enter into consultative and partnership arrangements to promote prevention?
☐ Are there features of the behaviour that qualify intent and traumatic impact? (E.g. is the behaviour: malicious; unintended; normalized in the work culture; or subject to perceptual differences?)	☐ Are guidelines for coping and case-examples provided?	
	Advice	☐ Are case examples of the positive contribution of preventive measures to business and society disseminated? (E.g. to productivity, health and well-being, participation, dignity at work, corporate social responsibility and sustainability, innovation, transparency, and corruption-prevention)
☐ Are there witnesses and can incidents be recorded?	☐ Is independent confidential advice provided, both from within the organization and external sources? (E.g. to account for situations where perpetrators are power holders)	
☐ What is the opinion of trusted co-workers, friends, family or independent advisors?	☐ Are recipients orientated to response options, employment assistance, and conflict mediation?	☐ Do industry and governmental agencies communicate positive case illustrations of the contribution of the cross-sectoral approach to safety and security to vitality and viability in local communities and local,
☐ Does the employer's policy framework address the behaviours, and is confidential advice available?	*Staff development*	
☐ Are there breaches of criminal law or other regulations, and what are the consequences if I report the breaches? (E.g. OSH, equity, harassment or employment laws)	☐ Are coping and resilience capacities developed? (E.g. skills in assertion, communication, conflict resolution, and stress management)	

Table 9.3b **Audit of capacity to promote constructive coping, zero tolerance policies and regulatory obligations – continued**

Individual capacity	Organizational capacity	Governmental capacity
☐ What are the costs and benefits of complaining/not complaining, and how can I cope in either case?	*Monitoring* ☐ Are risks of cash or 'hot/risky' products identified?	regional and state reputations? (E.g. in terms of jobs, wealth creation, security, and the quality of life)
☐ Has my personal risk-taking, provocation or vulnerability contributed to the incident and how can I be self-responsible?	*Vitality and viability enhancement* ☐ Are safe environmental design and emotional management strategies used to reduce stress/aggravation? (E.g.: work/life policies; child care provision; clean-green convivial office design; exercise programs; head and shoulders massage; recognition and rewarding of staff; tension-reducing community engagement)	*Obligations in compliance standards and guidelines* ☐ Are employers, employees and members of the public regularly reminded of regulatory obligations to prevent bullying/violence and their contribution to health and safety, productivity, competitiveness and the quality of work and community life?
☐ How can I be assertive, maintain a positive outlook, and sustain my health and safety?		
☐ What are my response options and what is the best option, in weighing up the risks?	*Research* ☐ Is research conducted into positive and negative patterns of coping to inform staff development?	*Research partnership* ☐ Are industry-wide studies of bullying/violence initiated? ☐ Are sectoral responses scoped and benchmarked against best-practice?

Table 9.3c **Audit of capacity to resolve complaints/grievances informally and formally and to reduce absenteeism and workers' compensation claims**

Individual capacity	Organizational capacity	Governmental capacity
Establish support network ☐ What are the opinions of friends, workmates, the union representative, the EAP, medical advisors, counsellors, and support groups? *Complete complaint/grievance risk assessment* ☐ Are the organization's procedures up to date, accessible, workable and compliant with regulatory standards or guidelines? ☐ Are complainants subject to retribution or stigmatization? ☐ What are my goals in pursuing the complaints/grievances process? ☐ What is the likely cost in monetary, emotional and career terms? ☐ Can the conflict be safely resolved in face-to-face discussion with perpetrator/s? (E.g. inform the perpetrator/s that their conduct is unacceptable and a formal complaint will ensue if the behaviour is repeated) ☐ Could I request a supervisor, manager or third party to speak to the perpetrator?	*Mainstream conflict resolution and restorative justice in the business strategy* ☐ Are skills for informal conflict resolution on the job developed and rewarded? ☐ Are managers and supervisors empowered to resolve complaints promptly, on the job? ☐ Are complainants obliged to complete a complaints/grievances risk assessment? ☐ Does conflict mediation start with 'no-blame' procedures consistent with principles of transformative justice, with provision for escalating penalties for serial offenders? *Data collection and reporting* ☐ Are *actual* complaints, complaint outcomes, and exit interviews recorded and reported? ☐ Are measures of prevalence, impacts, and outcomes taken to assess *likely* incidence and cost? (E.g. in a bi-annual survey?)	*Obligations in regulations, standards and guidelines* ☐ Do regulations/guidelines oblige the appointment of contact and mediation officers, and record keeping and reporting? ☐ Do industry associations and regulatory bodies require local independent conflict mediation before complainants can access grievance proceedings? ☐ Are complainants and alleged perpetrators supported in complaints/grievance processes, with provision for representation made available to those suffering psycho-physical injury? *Protection of complainants* ☐ Are complainants who pursue due process in complaint/grievance procedures protected against retribution and defamation charges? ☐ Is protection for whistle-blowers respected where complaints are made in the public interest, and is bullying/violence to them prevented?

Table 9.3c Audit of capacity to resolve complaints/grievances informally and formally and to reduce absenteeism and workers' compensation claims – *continued*

Individual capacity	Organizational capacity	Governmental capacity
Maintain health in complaints/grievance process ☐ How can I stay healthy, given complaints/grievances can be of indeterminate cost and duration due to perceptual, witnessing, and procedural difficulties? ☐ Can self-help assist me to maintain my health, esteem, and positive outlook? (E.g.: learning about bullying; meditation; relaxation; leisure; personal development; assertive training; anger management; dealing with difficult people; counselling; and/or life coaching)	☐ Are *actual* recorded complaints reconciled with *likely* incidence/cost derived from the work environment survey? ☐ Are experiences in complaints/grievance processes subjected to case analysis? *Valuing of outcomes* ☐ Are case examples of the positive contribution of prevention to organizational futures communicated? (E.g. the contribution to: productivity; relations with employees, clients, suppliers and contractors; sustainability; and community enhancement) ☐ Are employees who make positive contributions to the resolution of complaints and grievances acknowledged and rewarded?	*Recording and reporting* ☐ Do OSH, workers' compensation, human rights, and industrial authorities mandate recording of incidents/costs? ☐ Are workers' compensation costs compared across sectors? *Research partnership* ☐ Are complaint/grievance procedures in industry sectors scoped and evaluated against best-practice? ☐ Are causes of deteriorating health of victims in complaint/grievance processes subject to investigation as a factor that compounds the severity of impacts of original incidents?

Table 9.3d Audit of capacity to prevent the escalation of risk/severity to the point legal claims are launched and the victim is forced out of the workplace

Individual capacity	Organizational capacity	Governmental capacity
Assessment of risks of undertaking legal action	*Legal risks where a victim is unable to return to work or is constructively dismissed*	*External employment services agency support*
☐ Is independent advice about legal options available?	☐ What legal risks does the organization face? (E.g.: unfair/constructive dismissal; breach of terms in the employment contract; human rights/anti-discrimination commission; Ombudsman; Administrative Appeals Tribunal; damages claim for negligence/vicarious liability; and/or crime and misconduct commission)	☐ Is external agency advice and support concerning legal options available for persons who are: precariously employed; in workplaces lacking policies; and where management, owners/directors or union representatives are perpetrators?
☐ What are the risks, costs, durations and likely outcomes of legal pathways in my case?		☐ Does the external agency advice and support encompass: counselling and referral services? (E.g. to self-help, support groups, medical officers, psychiatrists, counsellors, alternative therapies, mediation, alternative dispute resolution, and lawyers)
☐ Realistically, which legal pathway is most appropriate, given: the causes, nature and severity of my injury; degree of PTSD; the quality of my evidence; and my resources and coping capacity?	☐ What are the likely costs of legal action? (E.g.: time of organizational staff; loss of productivity; undermining of reputation; payments to lawyers; and settlement of claims or judgement)	
☐ Could I claim whistleblower protection in respect of public interest disclosure, and what would be the costs and consequences?	*Accountability*	*External conflict mediation*
☐ Is support for legal action available (e.g. from the union, or an employee services agency)?	☐ Are costs of the deterioration of unresolved complaints/grievances into legal claims recorded?	☐ Are independent mediation and dispute resolution services that provide alternatives to legal action available to victims who are unable to resolve problems at work?
☐ On what basis can I select a lawyer? (E.g. agency recommendation, client experiences)	☐ Are costs of complaints that deteriorate into legal claims charged against departmental profits and hence management bonuses?	☐ Is there an Ombudsman empowered to resolve complaints about bullying/violence that are unable to be resolved in the workplace?

Table 9.3d Audit of capacity to prevent the escalation of risk/severity to the point legal claims are launched and the victim is forced out of the workplace – *continued*

Individual capacity	Organizational capacity	Governmental capacity
Written agreement with legal representatives ☐ Can the basis of all fees and charges be stated? (E.g. the basis of charging for the time of law clerks, lawyers, barristers, and incidentals) ☐ Has a time-line for the legal action been agreed? ☐ Will appropriate resources be committed to the action? ☐ Is my liability for residual or unexpected costs clear in a no-win, no-fee arrangement? (E.g. what if the settlement/award of damages is less than case costs?) ☐ Can provision for independent mediation of any conflict with my legal representatives be made?	(E.g.: absenteeism, staff turnover; time corporate officers and consultants take in sorting out problems; and legal costs) *Reducing legal risks* ☐ Can principles of restorative justice be applied to resolve angst, resentments, mistrust and grudges, and help parties move on from confrontation to reconciliation? ☐ Are managers empowered to apologise and reach settlements of claims acceptable to all parties, with due regard to precedent?	*Certification and rewarding of best practice* ☐ Is organizational best-practice certified by public and private agencies? ☐ Are compliant corporations and industry sectors acknowledged and rewarded? *Research partnership* ☐ Are costs of bullying/violence estimated at industry sector, region, state and national level to justify investment in prevention?

Table 9.3e Audit of capacity to manage the transition from workers' compensation to rehabilitation/return to work with ongoing attention to PTSD symptoms

Individual capacity	Organizational capacity	Governmental capacity
Facing the realities of surviving in workplaces under pressure	*Rehabilitation and return to work*	*Regulatory obligations*
☐ What vulnerabilities are likely to surface on my return to work?	☐ Is there compulsory assessment of psychological and physical rehabilitation needs?	☐ Do workers' compensation regulations oblige rehabilitation that addresses both the psychological and physical dimension of injuries produced by bullying/violence?
☐ Can I access rehabilitation that deals with both the psychological and physical trauma of bullying/violence and builds coping that enables my return to work?	☐ Do the rehabilitation and return to work planning processes include consultation with the victim, their advisors and family members?	☐ Is preparation of a return to work plan agreeable to the victim and the organization obliged?
☐ Are unrealistic expectations of benevolence, justice, and lack of pressures and conflict in contemporary workplaces impeding my return to work? (E.g. given most experience serial restructuring and work intensification)	☐ Are outstanding issues and concerns resolved to the satisfaction of the victim prior to their return to work? (E.g. to achieve restorative justice and reconciliation)	☐ Is the return to work process subjected to independent monitoring and review?
☐ What expectations and behaviours do I need to change to be more attuned to the realities of the workplace, and how can I change them? (E.g. cognitive behavioural therapy)	☐ Is the plan sensitive to problems of returning a victim to a work group or location in which perpetrators continue to behave inappropriately without sanction?	☐ Are appropriate job termination and workers' compensation procedures obliged where the victim is unable to return to work?
☐ What survival skills do I need *if* the realities are that bullying/violence persists in my workplace without effective interventions or sanctions?	☐ Are locations in the organization, tasks, workloads, developmental activities and supervision included in progressive steps in the return to work program?	☐ Is independent support and advice available to persons who appeal against dismissal of workers' compensation claims?
	☐ Can part-time work and homework be factored into the plan, if it assists the victim?	☐ Is an independent Ombudsman empowered and resourced to resolve claims and appeals that remain undetermined for an unreasonable time period?

Table 9.3e Audit of capacity to manage the transition from workers' compensation to rehabilitation/return to work with ongoing attention to PTSD symptoms – *continued*

Individual capacity	Organizational capacity	Governmental capacity
☐ Can the lack of effective policies and procedures be made an issue? (E.g. with the union, HR, the health and safety officer)	*Monitoring* ☐ Is there provision for monitoring of progress in return to work and resolution of difficulties?	*Data collection and reporting* ☐ Are workers' compensation and rehabilitation data for the impacts of bullying/violence collected by insurers and collated and reported for purposes of informing prevention?
Consider options outside the organization ☐ If I am unable to return to the workplace, or it is unsafe, what other work options are there?	*Data collection* ☐ Is case data collected, analysed and reported to improve management of rehabilitation and return to work?	*Rewarding* ☐ Are recognition and rewards accorded to compliant organizations and industry sectors? (E.g. reductions in workers' compensation premiums)
☐ Are there skills training opportunities in other career pathways that interest me?	*Accountability* ☐ Are workers' compensation, rehabilitation and return to work costs attributable to bullying/violence charged to departmental profits? (E.g. workers' compensation premiums, and rehabilitation and return to work costs?)	*Research partnership* ☐ Do government agencies and industry associations initiate research into causes and costs of workers' compensation claims and rehabilitation to inform prevention?
☐ Is down-shifting to a work/life arrangement that is less stressful or less conflict prone possible?		
☐ Are there job opportunities in other workplaces that are less stressful or conflict prone?		

Table 9.3f Audit of capacity to move on from legal action, suicidal ideation and revenge seeking, to learning, re-affirmation, reconciliation and constructive re-engagement with work and life

Individual capacity	Organizational capacity	Governmental capacity
Surviving legal action for damages ☐ Against what criteria should I decide to initiate, settle or discontinue legal actions for large damages claims for negligence/vicarious liability? (E.g. weigh up risks in terms of duration, costs, stresses, and likely outcomes)	*Constructive interaction with victims who have unresolved claims for damages* ☐ Are victims with unresolved claims treated respectfully? ☐ Is support for counselling and medical services continued? ☐ Are employee services maintained? (E.g. access to e-mail, child-care, social, health and vacation club, discount purchase plans)	*Commitment to international laws/ conventions* ☐ Is there national government support for international laws and conventions that prohibit bullying/violence within workplaces and in corporate operating environments? (E.g.: ILO conventions; Human Rights; the International Criminal Court; UN resolutions mandating interventions to prevent genocide, terrorism, and the degradation of local economies and communities)
Preventing extreme reactions ☐ Where can suicide prevention help be obtained to deal with suicidal feelings? ☐ How can I move beyond overwhelming desires to take revenge against perpetrators, employers and elements of the system that have failed me? (E.g. more healthy: collective political action for reform; anger/ hate management; psycho-therapy; and forgiveness and reconciliation)	*Reducing risks of large damages claims for negligence/vicarious liability?* ☐ Is alternative dispute resolution acceptable? ☐ Can claims be settled early on the basis of an independent 'reasonable person' test? ☐ Does the legal defence displace responsibility to the victim and insurers, and leave risky systems, work practices, and behaviours unexamined?	*Mainstreaming prevention within competitiveness* ☐ Does the government orchestrate partnerships to address bullying/ violence that enhance market opportunities? (E.g. through increased vitality, viability, safety and security, and employment, in local communities?)

Table 9.3f Audit of capacity to move on from legal action, suicidal ideation and revenge seeking, to learning, re-affirmation, reconciliation and constructive re-engagement with work and life – *continued*

Individual capacity	Organizational capacity	Governmental capacity
Moving from victim to survivor	*Mainstream prevention of bullying/violence in the business strategy*	☐ Are there conjoint government, industry, union and community certification, recognition and rewarding of best-practice prevention? (E.g. are compliant organizations rewarded in public awards and in trading, contracting, tendering and licensing arrangements)
☐ What interventions might help me move on from the situation, and obsessive desires for justice, revenge or self-harm? (E.g. grieving, shame management, cognitive behavioural therapy, self-affirmation, forgiveness, reconciliation)	☐ Does the organization learn from outcomes of the evaluation of the anti-bullying/violence policies and procedures?	
☐ How can I reaffirm my self-esteem, dignity, and integrity?	☐ Do the anti-bullying/violence measures enhance the business strategy? (E.g. through staffing, resource allocation, decision-making, product/market, investor and community relations free of bullying/violence)	☐ Are productive and societal costs modelled from scientific evidence on an industry/state/national basis to inform the cross-sectoral preventive strategy? (E.g. are costs of bullying/violence calculated in terms of welfare, medical expenses, foregone opportunities and innovations, suppression of intelligence, and corruption?)
☐ How can I compose a new positive life narrative as a basis for moving on? (E.g. therapies and coaching that moves from destructive to healthy life narratives)	☐ Are the overall costs of bullying/violence to the organization estimated on the basis of scientific research?	
☐ What positive things have I learned from my experience, and how can I apply my learning in new work or social situations?	☐ Are outcomes of the anti-bullying/violence policy expressed in social accountability and responsibility (E.g. in terms of the triple bottom line accounting)?	☐ Do government initiatives legitimate provision of development of market clusters involving industry, employee, government and third-sector organizations to address problems of bullying/violence to mutual benefit? (E.g. in smart state, clever country initiatives?)
☐ Could my learning make a positive contribution to work and society? (E.g. writing my survival story; engaging in awareness-raising, victim-support, advocacy, and policy development)		

Conclusion

The cross-sectoral web of safeguards proposed in this chapter has been aimed at preventing the intensification of risk and severity of bullying/violence due to the interplay of individual, organizational and societal risk factors (Figures 9.1 and 9.2). 'Intensities' in which the risk and severity of bullying and violence commonly escalate have also been modelled. These intensities commonly form in the interplay of vulnerability, global forces, destruction of psychological integrity, suicide ideation, litigiousness, the hardening of borders, and revenge.

A case analysis of legal actions in respect bullying/violence has indicated that recipients commonly pursue remedies over long time periods and bear a growing burden of interactions, costs and added duress that can increase the severity of impacts (Table 9.1). Critical stages at which risk and severity escalate have been identified in the modelling of typical 'risk/severity spirals' (Figures 9.3 and 9.4). Recipients were commonly proactive in seeking help from a variety of personal and professional sources within and without the organization. The quality of this help had the potential to mediate the escalation of risk and severity of incidents (Table 9.2).

Risks that the severity of impacts will escalate are high where recipients have less developed coping and resilience and abrasive interactions are normalized. Where employees and managers, as well as clients, contractors, and suppliers, are not obliged to responsible risk management by organizational codes of conduct and governmental compliance standards, there is little to constrain the escalation of risk and severity. Many employers continue to displace costs of bullying/violence to their insurers. Governments are unlikely to have the political courage to implement compliance codes that oblige intervention unless concern about the risks of workplace bullying/violence is widespread in the community and shared by employee representatives and management.

Government leadership of partnerships to develop codes that both oblige and reward the implementation of preventive strategies is considered necessary to promote commitment to safeguarding across the sectors. Incentives to implement preventive policies and procedures need to be generated through conjoint regulatory, contractual and market agreements that encompass industry, employee, professional, third sector and institutional interests. Audit tools presented in Tables 9.3a–f have been designed to assess needs to build capacities of key stakeholders to participate in a cross-sectoral web of safeguards that realizes mutual benefits from the enhancement of skills, productivity, health, safety, transparency and competitiveness.

10

Towards a Global Response to Workplace Bullying and Violence

Paul McCarthy

This concluding chapter uses evidence presented in preceding chapters as the basis for the inauguration a global initiative to prevent workplace bullying/violence. Presently, preventive measures within national and regional/state jurisdictions are localized, mainly in wealthy industrialized countries, and are fragmented across OSH, employment relations, legal and criminological responses. Lack of widespread application of procedural referents contributes to difficulties in safeguarding. The development of policies and strategies that have widespread acceptance in the international community will be at the core of a global initiative. A mutually supportive balance of regulatory responses and organizational value-adding approaches geared to indicators of sustainability is considered desirable to optimize safeguarding. A coordinated international response requires commitment by international agencies, nation states, NGOs and TNCs to realize benefits for all stakeholders. A preliminary agenda for a global response is canvassed below.

The evolution and internationalization of safeguards

The discussion in this section examines the present pattern of the evolution of safeguarding as a guide to future improvements. The analytical framework depicted in Table 10.1 is drawn from Zadek's (2003) modeling of responses to salient social and environmental problems. This table shows the clustering and interaction of diverse interests in facing the challenge of workplace bullying/violence through service provision, preventive partnering, and public policy driven regulatory safeguards aligned to enhancement of competitiveness. In the application of this framework, the dimension of *internationalization* has been

Table 10.1 The clustering and interaction of interests in the evolution of responses to workplace bullying/violence

Clusters	Responses	Interaction of clusters in the evolution of responses
Awareness-raising and challenge	Media 'horror stories'; formation of victim support groups; Publication of research findings; Networking of interests in advocacy for government intervention	Stories of victimization circulated in the media are legitimated by expert commentary attributing experiences to 'bullying' and 'occupational violence'. Demonization of perpetrators, moral panic, advocacy for victim's rights, provision of services by professionals, research studies, and calls for punitive legal sanctions all ensue. Governments respond through consultative policy development that generates further awareness and airing of demands for protection.
Market-making and service provision	Market-based services provided by support groups, publishers, EAPs, medical officers, psychiatrists, counsellors, lawyers, unions, consultants, and researchers	Services to victims and employers legitimate the attribution of experiences to 'bullying' and 'violence'. Self help books generate awareness of response options. Expert witnesses inform legal actions. Consultants service corporations and governments reacting to the challenge. Researchers seek funding for studies to inform responses. The implementation of risk-management strategies is accompanied by bullying/violence awareness training. A bullying/violence industry emerges.
Partnerships for prevention	Formation of partnerships involving employee, professional, community, governmental and institutions to realize mutual benefits in responding to the challenge	Interest group advocacy, evidence of costs, and difficulties in responding stimulate partnering between sectoral interests in the development of responses. Research partnerships between industry, governmental and academic interests are formed to assess needs and inform policy development. Employee, professional, industry and governmental interests form partnerships that develop self-regulative guidelines. Self-regulative responses produced by cross-sectoral partnerships remain fragmented, difficult to access and prone to failure. The call for regulatory intervention by victim support groups, employee representatives, and helping professionals continues.

Table 10.1 **The clustering and interaction of interests in the evolution of responses to workplace bullying/violence –** *continued*

Clusters	Responses	Interaction of clusters in the evolution of responses
Regulations and obligations stemming from public policy responses	Governmental leadership in the development of self-regulative responses through consultative processes. Orientation and extension of the existing regulatory framework to reduce the risks.	The governmental response is oriented to self-regulative, third-way solutions in the interests of effective resource use and reciprocal obligations that realize benefits for all stakeholders. Advocacy for regulatory interventions continues due to persistent difficulties in accessing remedies and lack of preventive efforts by business. Departments of OSH and employment relations lead the development of codes of practice and advisory standards that oblige organizations to implement risk management practices, including prevention, grievance procedures, recording, reporting, training and rehabilitation.
Internationalization of response	Advocacy of global policy responses through the networking of international agencies and national interests in the development of a global response.	An incipient international response emerges from international exchanges between victim support groups, researchers and agencies via global information networks. Information is exchanged in international research forums in which positions are debated and shared understandings developed. The fragmentation of responses within nation states and international agencies stimulates the clustering of interests to further the development of international best-practice referents.

separated out to highlight the manner in which regional/national responses stimulate a clustering of transnational interests around global initiatives. The pattern of the evolution of responses is seen as less a linear progression to higher-order responses, and more a mutually reflexive process in which the complexity of responses increases due to interactions within and between clusters.

Regulatory responses in a global context

The review of regulatory responses that follows commences with a brief profile of the principles contained within a range of statutes and an assessment of their application to workplace bullying/violence (see: Table 10.2). Statutes within particular national jurisdictions are not dealt with in detail (for such applications see: Yamada, 2003; Report of the Queensland Workplace Bullying Taskforce, 2002; Wirth, 2002; Income Data Services, 1996). Rather, the discussion is concerned with ways of improving the scope and viability of different regulatory pathways and with the alignment of national regulatory frameworks to a global initiative to prevent workplace bullying/violence.

Limitations of statutory responses

The application of statutory principles to workplace bullying/violence was shown in Table 10.2 and in previous chapters. These discussions identified the following limitations:

- Responses remain fragmented within and between national jurisdictions and pose difficulties in accessing remedies. This fragmentation also impedes consistent responses to forms of workplace bullying/ violence stemming from globalization that are transmitted by TNCs and international agencies.
- Difficulties in defining and evidencing less overt forms of workplace bullying/violence need to be addressed in viable statutory responses.
- Presently, the pursuit of statutory remedies post-event can draw plaintiffs into actions of indeterminate cost and duration, and added duress. Thus, failure to resolve issues in the workplace can lead to the displacement of costs onto victims, insurers and the wider society.
- The attribution of bullying/violence to the personality of the victim and the perpetrator can occlude consideration of duty of care requirements enshrined in OSH legislation.
- Determining whether a less overt negative action constitutes bullying/violence can be 'undecidable' (Catley and Jones, 2001).

Table 10.2 **Statutory remedies for workplace bullying/violence**

Statutes	Application to workplace bullying/violence
Criminal laws	Prohibit homicide, assaults, unwanted physical contact, threats of violence, and stalking that can occur in workplaces. Police may be reluctant to intervene in workplace conflicts, pranks or initiation rituals, although these can escalate into more serious incidents. International criminal laws may be applicable to the organized perpetration of crimes against humanity that can have devastating impacts on workplaces and their environs.
OSH legislation	Employers are obliged to provide safe work-places/processes. The emergence of research literature, scientific evidence and guidelines in most western countries (see: Einarsen et al, 2003; Di Martino et al, 2003) has made forms of workplace bullying/violence definable, foreseeable, and preventable, and hence subject to OSH duty of care obligations. In some cases, definitions of workplace bullying/violence have been specified in specific subsidiary legislation, codes of practice, or advisory standards (e.g. DIR, 2004; WorkSafe Western Australia, 1999; Swedish NBOSH, 1993). OSH legislation requires employers to develop and implement preventive policies and procedures in consultation with employees and to provide for training, recording, and reporting (e.g. McCarthy et al, 2001a,b; 2002b).
Workers' compensation legislation	Compensation can be sought for physical and psychological injuries. However, claims are often dismissed on the basis that injuries may not be entirely attributable to workplace causes but to the claimant's prior psychological and physical pathologies and other life stresses. Appeals processes can be arduous, proceed over several months, and costly legal advice may be necessary. All the while, the claimant may unable to return to work and the added stresses of pursuing claims may worsen their health in spite of rehabilitation. Lawyers often recommend that victims pursue workers' compensation claims in the interests of generating evidence of injury to support common law actions against employers for substantial damages in respect of negligence. Excessive workers' compensation claims can also lead to increases in insurance premiums. Workers' compensation determinations can also be used as evidence of bullying/violence in staff performance appraisal, development and discipline. The use of workers' compensation findings in the workplace disciplinary processes can raise questions about rules of evidence and natural justice for perpetrators.

Table 10.2 Statutory remedies for workplace bullying/violence
– *continued*

Statutes	Application to workplace bullying/violence
Industrial law	Employment contracts commonly provide for a hierarchy of complaints, grievance and disciplinary procedures within workplaces. Where conflicts cannot be resolved within workplaces, disputes may be taken to industrial tribunals. Remedies for unfair or constructive dismissal may be sought in industrial tribunals where bullying/violence forced a person out of the workplace. Costly legal support may be necessary to improve chances of success in tribunals. The support of union representatives and legal advocates can facilitate applications to industrial tribunals. However, access to this support is limited by low levels of union membership in the new economy, particularly amongst the precariously employed. Lack of specific knowledge about bullying/violence among employee representatives, professionals, managers and victim support groups can limit the quality of advice and support.
Anti-discrimination legislation	The application of anti-discrimination legislation to workplace bullying/violence is limited to the perpetration of negative acts that can be considered discriminatory within key terms of the act (e.g. gender, age, marital status, religion, disability etc.). Where bullying/violence can be construed as a form of sexual harassment, then remedies can be pursued through anti-discrimination tribunals. Such tribunals provide plaintiffs with cost-effective remedies that stand as a model for statutory remedies for workplace bullying/ violence. However, governments appear to have avoided extending anti-discrimination legislation to address workplace bullying/violence or provision of an equivalent regulatory response due to funding concerns.
Administrative law	Workplace bullying/violence can be construed as a breach of public sector administrative laws with remedies sought in appeals tribunals. However, appeals can require costly legal support, place victims in conflict with power holders, and lead to their ostracization and treatment as whistleblowers.

Table 10.2 **Statutory remedies for workplace bullying/violence**
– continued

Statutes	Application to workplace bullying/violence
Public sector ethics acts	CEOs are mandated to implement policies and procedures that ensure employees, clients and members of the public are treated with respect, dignity and integrity. Workplace bullying/violence infringes ethical prescriptions. Thus, behaviours constituting bullying/violence can be listed in organizational codes of conduct and remedies sought through related grievance and disciplinary procedures. Ethical violations involving workplace bullying/violence may also be referred to integrity or crime and misconduct commissions.
Dignity at work legislation	A private member's Dignity at Work Bill developed with union support is currently proceeding through readings in the British parliament. Such legislation can express definitional terms, obligations, and aspirational standards that are useful in raising the public profile of the issue and to legitimate and orientate prevention.
Common law	Action for damages to compensate for physical and psychological injuries due to workplace bullying/violence may be taken against a perpetrator an/or their employer. Redress from employers may be sought in terms of vicarious liability arising from negligent behaviours that constitute a failure to protect a person from physical or psychological injuries (see: State of New South Wales v Seedsman [2000] NSWCA). Damages may also be sought for breaches of implied contractual terms that entitle parties to perform their side of the contract in peace, i.e. without being subjected to bullying and violence. Unfortunately this pathway can engage complainants in legal actions of indeterminate cost and duration, and added duress. Favourable outcomes in one court may have to survive appeals processes, and no-win no-fee arrangements do not always protect plaintiffs from financial costs.
Ombudsmen	The empowerment of ombudsmen to intervene in workplace bullying/violence can provide a cost-effective remedy for victims who are unable to gain redress within organizations or other external pathways (e.g. Office of the Employee Ombudsman, 2000).
Public interest disclosure legislation	Whistleblower protection legislation can safeguard employees against reprisal as a consequence of their making complaints about unethical or illegal behaviours. Bullying/violence is often used to enable corrupt/unlawful practices to persist, and to retaliate against whistleblowers. Whistleblowers commonly end up disadvantaged after making complaints and can be forced out of their jobs.

Table 10.2   **Statutory remedies for workplace bullying/violence
– *continued***

Statutes	Application to workplace bullying/violence
Trade practices legislation	Where workplace bullying/violence is perpetrated in commercial transactions, it may be construed as a breach of fair trading regulations. Such regulations conceivably empower trade practices commissioners to initiate punitive actions, hear complaints, and award damages in respect of trading practices involving workplace bullying/violence.
Consumer protection legislation	Remedies for unreasonable treatment of consumers involving forms of bullying/violence can be sought where consumers are maltreated during sales transactions and related financing and servicing arrangements. The provision of cooling off periods and disclosure of vested interests by vendors and sales organisations safeguard against such behaviours.
Financial service regulations	To the extent that governance standards are listed in financial services regulations, there is a potential for workplace bullying/violence to be construed as improper governance and a breach of regulations. Requirements for reporting in respect of sustainability indicators and triple bottom line accounting can also be specified in financial service regulations. Best-practice indicators of sustainability performance and social accountability can make reference to labour standards, for example, the Global Reporting Initiative (2002) that refers to ILO labour standards.
Emergency services legislation	Regulations for the provision of emergency services can oblige corporations to plan, train and maintain records about the risks stemming from forms of bullying/violence perpetrated against employees, contractors, clients and members of the public. Bullying/violence can be implicated in conflicts, disputes, hold ups, extortion threats, hostage taking, sabotage, kidnapping and terrorism (e.g. bio-terrorism, attacks on property, electronic systems and individuals) within organisational operating environments. Thus, emergency service legislation can obligate organizations to implement policies and environmental designs that safeguard against bullying/violence.

There are often difficulties in unequivocally deciding whether forms of bullying/violence are intentional, malicious, or unreasonable. Pranks, initiation rites, and normalized physical contact at work may be considered cultural traditions by some, or gross violations by others.

- Statutory obligations and compliance standards can be less effective if they foster minimum compliance mindsets (Cohen and Grace, 2003: 445).

Recommendations to improve the statutory response

The discussion of limitations above has lead to the following recommendations to improve the statutory response.

- A web of statutory constraints oriented to 'best-practice' international standards is desirable, since lowest common denominator guides are unlikely to be effective in preventing the multiplicity of forms of workplace bullying/violence.
- An international database to record workplace bullying/violence events could be established to inform the development of statutory initiatives. For example, programs such as the *United Nations Surveys on Crime Trends and Operations of the Criminal Justice System* (2004) and the *International Crime Victimisation Survey* (Barklay and Tavares, 2002) could be extended to record occupational violence and bullying.
- The development of subsidiary legislation in respect of bullying/ violence within OSH statutes could be undertaken to provide more stringent compliance standards.
- Enhanced training and resourcing of OSH inspectors to identify breaches of OSH duty of care and powers to issue provisional improvement notices in respect of workplace bullying/violence would further early preventive efforts.
- Both victims and alleged perpetrators should be entitled to natural justice. Complainants should also be protected from risks of defamation through procedural arrangements.
- Where possible, no-blame conflict mediation and transformative approaches should be applied, with provision for developmental options for both perpetrators and recipients, together with escalating sanctions for proven serial offenders. Difficulties in defining and evidencing less overt forms of workplace bullying/ violence and its normalization in work cultures justify this approach.

- In adjudicating complaints about workplace bullying/violence, the focus should be on the perpetrator's behaviour. The physical or psychological inadequacies of the recipient should not excuse perpetrator behaviours or work practices that are unlawful or unreasonable. Furthermore, unreasonable work practices that contribute to conflicts need consideration in adjudication of responsibility.
- Regulatory obligations for emergency services and environmental design should be aligned with the anti-workplace bullying/violence policies. Responses by emergency services (ambulance and fire), police (SWAT teams), and health authorities (testing for toxic chemicals, decontamination, and treatment of infectious diseases) should also be consistent.
- The cost of insurance protection premiums can be aligned to the quality of preventive strategies implemented to safeguard against workplace bullying/violence.
- Ombudsmen should be empowered to address complaints that are unable to be resolved within organizations, to provide pathways for mediation that do not impose unreasonable costs and distress on recipients.
- The protection of whistle-blowers needs to be strengthened within legal statutes to safeguard against the use of bullying/violence to enable corruption to persist.

Finally, legal compliance standards and capacity building responses that add value should not be treated as polar opposites. The conduct of business operations in national and global arenas is contingent upon statutory obligations that enable business operations to proceed. Management buy-in to regulatory compliance is likely to be enhanced where workplace bullying/violence is addressed through a balance of compliance standards and strategies aligned to value creation (see: Haines, 2003).

Internationalization of the statutory response

Some international codes, declarations, programs and tribunals that can protect against gross forms of bullying/violence that can arise within, impact on, and be perpetrated by organizations are listed below. Notably, most of the codes, declarations, programs and tribunals discussed below do not have purchase in domestic jurisdictions unless they have been passed into legislation by national parliaments.

- The ILO (2003a) code of practice provides a point of reference for the integration of legal terms, evidence on patterns, and remedies that are presently fragmented within and between national jurisdictions.
- International trade agreements commonly make reference to labour standards and thus can be aligned to the ILO (2003a) *Code of Practice* and implemented by nation states. In addition, Article XX in the General agreement on Tariffs and Trade allows for import bans for the protection of '...human, animal or plant life or health life or health' (see: Singer, 2002: 74). However, a history of failure by the WTO to ban the trading of goods on the basis that violence has been used in the process of their production has been observed (Singer, 2002: 70).
- The International Covenant on Civil and Political Rights (1984) (ratified by 151 States) does have legal force and prohibits '...torture or other cruel inhuman, or degrading treatment' (Article 7) (see: Amnesty International, 2003: 1).
- The International Commission of Jurists has called for a universal jurisdiction for crimes including piracy, slavery, war crimes, crimes against peace and humanity, genocide, and torture (see: Singer, 2002: 128).
- The statute for the International Criminal Court was established in 1998. It lists crimes against humanity, including widespread acts against civilian populations causing '...great suffering, or serious injury to body or to mental or physical health'. Notably, the International Commission on Intervention and State Sovereignty reduced the grounds for humanitarian intervention to large-scale loss of life or ethnic cleansing in 2000 (Singer, 2002: 139–141).
- Respect for domestic jurisdiction within the UN Charter limits its potential to intervene in the affairs of member states. However, enforcement measures may be enacted in respect of '...threats to the peace, breaches of the peace and acts of aggression' (Chapter V11 of the UN charter cited in Singer, 2002: 143). In the past, the UN has also established international tribunals to address crimes against humanity, for example, in the former Yugoslavia and Rwanda.
- The UN Code of Conduct for Law Enforcement Officials encompasses: '...protection against abuses, whether mental or physical' (Article 5) (see: Amnesty International, 2003: 1).

- The UN Basic Principles on the use of Force and Firearms by Law Enforcement Officials exhorts officers to: '...apply non-violent means before resorting to the use of force and firearms' (Article 4) and also to '...[m]inimize damage and injury, and respect and preserve human life' (Article 5). This principle also calls for the careful evaluation of risks from non-lethal incapacitating weapons to uninvolved persons and controls on usage (see: Amnesty International, 2003: 2).
- A resolution of the UN General Assembly in 2001 called on governments to prohibit the: '...production, trade, and export and use of equipment that is specifically designed to inflict torture or other cruel, inhuman or degrading treatment'. In addition, the European Commission has proposed a Council Trade Regulation that would ban trading in such equipment (see: Amnesty International, 2003: 3).
- A Declaration by the UN Conference on Environment and Development (1992) called on States to apply a *precautionary principle* in implementing protection against threats of serious or irreversible damage to the environment, including human health. The call was made on the basis that: '...lack of full scientific evidence shall not be used as a reason for postponing cost-effective measures' (see: Cohen and Grace, 2003: 453)

Notwithstanding their limited application in domestic jurisdictions, the international responses outlined above prompt awareness that calamitous loss of life, injury and degradation of the human condition can stem from workplace bullying/violence. The sources of regulatory obligations outlined above could be open to more specific recognition of ways that workplace bullying/violence are implicated in such gross violations.

Sustainable safeguarding

Further development of the international web of regulatory safeguards against workplace bullying/violence may be possible through linkage with the global sustainability movement. The discussion in this section examines the potential for a global anti-workplace/bullying initiative to create mutual benefits for stakeholders through enhancing organizational performance on indicators of sustainability.

Hogarth *et al* (2003: 374) have linked OSH and sustainability in claiming that: '... no company that presides over avoidable deaths, injuries and illnesses among its own people will ever be able to claim that its business is sustainable'. Their *safe companies framework* is premised on a sustainability-based business approach built on OHS principles and the empowering of safety, health and environment (SHE) professionals to play a more influential role in corporate life. These activities are projected across a 'safety-sustainability continuum' extending from provision of a safe workplace to 'fulfilling responsibility for projecting the planet and its people' (Hogarth *et al* 2003: 380).

The implementation of programs such as those of Hogarth *et al* will require significant capacity building amongst SHE professionals and others in their organizational chain of command. The challenge will be to move beyond meeting minimum legal standards to operationalizing principles of natural capitalism, bio-mimicry, and human rights in ways that conjointly address poverty, social and environmental degradation and the bullying/violence that may further it (Suzuki and Dressel, 2002; Hawkens *et al*, 1999).

The logic of the public reporting of sustainable performance is that it enhances the reputational capital of organizations and with that improves business prospects, investor regard, and attitudes of regulators (Grossman, 2003; McCarthy *et al*, 2002b, 2003b). The following definition of reputation (Grossman, 2003: 431) indicates a number of points at which workplace bullying/violence might corrode the regard of a corporation by diverse stakeholders.

> ...[R]eputation might be best defined as the distinguishing point that defines how an organization is recognized by its internal and external stakeholders for its capacity to maintain the esteem and integrity of its employees, control the quality of its services or products, minimize its environmental footprint, deliver sound financial performance, maintain a beneficial social impact and demonstrate an open transparent approach to public reporting and corporate governance.   (Grossman, 2003: 431)

Prospects for reputation to leverage the implementation of sustainable strategies within organizations have grown with acceleration of information flows in the new global economy. Corporations that perpetrate brutal work practices can now quickly find themselves subjected to a range of actions, including protests and consumer boycotts (Watts and Lord Holme, 1998) that can impact adversely on their market performance and operating licenses. Stakeholder activism has in some cases lead TNCs to address inhumane work practices (e.g. Nike, see:

Community Aid Abroad, 2000). Allegations of workplace bullying/ violence are also likely to arise in commissions of inquiry and legal actions contesting the responsibility of corporate activities (see: Romei, 2002).

Progress in the mainstreaming of sustainability within management practice is occurring, albeit slowly. A Copenhagen Centre (2002: 7) report has indicated that less than 3,000 major companies out of some 60,000 TNCs (which have some 800,000 foreign affiliates) appear to be '...systematically and strategically addressing corporate social respons- ibility issues throughout their operations'. Factors contributing to progress include the rise of environmental consciousness due to the circulation of information through global networks, increasing stake- holder activism, the growing use of ethical investment and social accountability rating systems, and triple bottom line accounting.

On the other hand, reputational cyberspace remains hotly contested as organizations deploy propaganda (see: Anon, 1999: 18–19) and manoeuvre to become less transparent (Grossman, 2003). Govern- ments also more often respond through third-way political strategies operationalized through cross-sectoral partnerships aligned to sustain- ability indicators (McCarthy *et al*, 2004). The increased use of commer- cial-in-confidence agreements and withholding of sensitive documents from freedom of information access tends to hide corporate and gov- ernmental processes. Corporations can also quickly use threats of defamation to silence allegations of inappropriate conduct (ABC, 2001).

The future strength and influence of a global anti-workplace bully- ing/violence initiative will be enhanced if a critical mass of interna- tional agencies, TNCs, NGOs and national governments subscribe to indicators of sustainability aligned to international best-practice for the prevention of workplace bullying/violence that realizes benefits for all stakeholders. In turn, mutual productive, market, social and political benefits of safeguarding can be assured by the pursuit of international best practice in recording, reporting and auditing of sustainability per- formance. In these respects, the Global Reporting Initiative (GRI) is a leading international source of best-practice indicators for the report- ing of organizational sustainability.

The GRI was established in 1997 under sponsorship from the UN Environment Program and the US non-governmental Coalition for Environmentally Responsible Economies (CERES). The GRI (2002) *Sustainability Reporting Guidelines* encompass a wide spectrum of eco- nomic, environmental and social criteria. Currently, the GRI guidelines do not directly acknowledge the possibility that bullying/violence might degrade organizational performance against sustainability criteria. However, the GRI does cross-reference to ILO codes, and could be expected to refer to the ILO (2003a) *Code of Practice* when finalized.

The recognition of bullying/violence at work as significant, mostly unaccounted, degraders of organizational sustainability (McCarthy *et al*, 2000b, 2003b) builds a strong case for indicators of safeguarding to be specified more directly in GRI criteria. For example, the GRI reporting framework could include criteria to gauge organizational performance in safeguarding against impacts of bullying/violence on OSH, employee rights, labour/management relations, and diversity. Indicators to prevent the bullying/violence that restricts whistle-blowing and enhances capacities for unethical practices (e.g. corruption, discrimination, child labour, forced and compulsory labour, and abuses of indigenous rights) could also be expressed in GRI criteria for *human rights*. GRI criteria for *strategic management, security practices* and *collective bargaining* could also directly acknowledge the implication of workplace bullying/violence in degrading sustainability performance.

In turn, the ILO Code (2003a) could also acknowledge the negative impacts of workplace bullying/violence on sustainability, which might promote linkage with the GRI and other social responsibility accounting and auditing systems. Indicators for safeguarding against the impacts of bullying and violence on social, economic and environmental performance could also be inscribed directly in indicators such as: *ISO 14,000; AA1000 Standard (ISEA, 2000); AccountAbility Standard 1000; Social Venture Network* (1999); and the *Dow Jones Sustainability Index*.

The GRI (2002) principles of best practice in reporting require transparency, inclusiveness, auditability, completeness, relevance, sustainability content, accuracy, neutrality, comparability, and timeliness. These standards provide a useful reference for assessing the reporting of performance in safeguarding against workplace bullying/violence. Thus, the GRI reporting principles could well be linked with the ILO (2003a) *Code of Practice* in the interests of a whole-of-UN agency response.

The case for a global anti-workplace bullying/violence initiative

The case for a global response needs to be made convincingly to international agencies, TNCs, NGOs and within nation states to generate a critical mass of support to enable the project to proceed. Some potential benefits for stakeholders are noted below.

Benefits for international agencies

Arguably, benefits to international agencies could accrue from a global initiative that:

- Provides referents for the implementation of policy and procedures to reduce threats of bullying/violence to agency operations, including adverse impacts on reputation, misallocation of resources, furtherance of corruption, and program failures;
- Enhances powers, consistent with agency charters, to reduce workplace bullying/violence that furthers cycles of poverty, ignorance, oppression, exploitation, corruption, misery, dominance by economic minorities, authoritarian governance, military spending, revenge, terrorism, war and genocide (see: Singer, 2002: 116–120; Chua, 2003); and
- Establishes policy referents to prevent systemic violence against less powerful nations and communities through global protocols on trade, development and finance (e.g. through safeguarding against large-scale capital intensive projects that: destroy local communities and inflict catastrophic environmental damage; restrict pharmaceutical patents; compound indebtedness; and promulgate unfair trade practices).

Benefits for nation state governments

Governments that subscribe to a global anti-workplace bullying/violence project could derive benefits from:

- Demonstration of caring about distress due to bullying/violence that has been experienced widely in electorates in recent years;
- Access to models of international best practice that legitimate and facilitate the implementation of consistent preventive responses within national jurisdictions;
- Enhancement of business and community perceptions of government leadership in cross-sectoral initiatives that build people management capacities;
- Optimization of conditions for 'responsible competitiveness' (Zadek *et al*, 2003) through promoting more vital and viable relations between management, customers, employees, contractors, suppliers, partners, investors, communities and regulators;

- Contribution to *clever-country* agendas through assuring work practices and governance that foster creativity and innovation and attract and retain highly skilled staff; and
- Furthering of *safe-community* programs that promote the nation and its regions as safe and viable places to live, work, do business, and visit.

Benefits for TNCs

Potentially, TNCs that implement zero-tolerance policies to safeguard against bullying/violence through international best-practice work environments and processes could:

- Reduce the costs of bullying/violence in their operations (see: Chapter 3);
- Secure the corporation against the use of bullying/violence to further corruption;
- Send a positive signal to clients, investors, international agencies, national governments and communities that the corporation takes care in pursuing sustainable economic, social and environmental practices;
- Justify the corporation's license to operate in national and regional/ state jurisdictions;
- Provide referents for debates with interests in third world and developing countries that contest the legitimacy of humanitarian work practices and regard their implementation as an imposition of trade barriers by wealthy western states;
- Establish level playing fields conducive to the realization of mutual benefits for players who abide by regulations, so that corporations that derive a short term benefit from bullying/violence are exposed to the scrutiny of stakeholders via sustainability reporting; and
- Protect against unreasonable and unconscionable behaviours by corporate officials that may result in poverty, social disruption, environmental damage, degraded safety and security, political instability, crime, violence and terrorism (see: Chapters 4 and 5).

Benefits for NGOs

The case for the commitment of NGOs to a global anti-workplace bullying/violence project can be made in terms of benefits that accrue through:

- Access to policies and strategies that can safeguard against involvement of the NGO and its partners in maltreatment of employees,

clients or community members, with consequent damaging revela-
tions and loss of support by subscribers and sponsors;
- Protection of NGO program operations from negative impacts of
  bullying/violence on the productivity and OSH of their employees;
- Provision of standards that enhance the development of people
  management skills and work practices within the NGO, and
  improve relations with supporters, clients, contractors, suppliers,
  partners and governmental stake-holders;
- Demonstration that principles of human rights, equity, justice and
  democracy (as are usually inherent in NGO charters) have been
  mainstreamed in the NGO's own practices via the implementation
  of the anti-bullying/violence policy and procedures;
- Legitimation of requirements that local partners implement best-
  practice anti-workplace bullying/violence guidelines; and
- Prevention of the use of workplace bullying/violence to further corruption that can be endemic in the external operating environments of some NGOs.

Agenda for a global response to workplace bullying/violence

Evidence presented in the previous chapters indicates that workplace
bullying and violence are pervasive interactive phenomena that arise
from interplay of local and global forces. As such, bullying/violence
in work environments have been identified as significant corrosives
of productivity, health and well being, reputation, transparency,
human rights, and of social and natural ecologies. The present fragmentation of responses within national jurisdictions, and the lack of
leverage against perpetration of workplace bullying/violence in
TNCs and international agencies, justifies the call for an organized
global response.

A key outcome sought from the agenda set out below will be the
generation of a critical mass of support in the international community to further advocacy for a UN Declaration of a global program
against workplace bullying/violence. Agenda items for action to
achieve a top-down whole-of-UN agency response are canvassed as
follows.

- Initiation of a global network to bring together diverse international
  interests in the finalizing an international code of practice for workplace bullying and violence that has widespread endorsement in the
  international community.

- Establishment of an international research forum to:
  - initiate research into comparative national incidence and severity estimates that can be entered into an international data base; and
  - study the interrelationships between workplace bullying/violence, sustainability, corruption, human rights abuses, and crimes against humanity, including terrorism.
- Making the case for international agencies, NGOs, national governments and TNCs to participate in development of a global workplace bullying/violence protocol geared to the realization of mutual benefits from the enhancement of productivity, OSH, social, environmental and political conditions, and competitiveness;
- Agreement on international best-practice terms of prevention for definition, policies, and strategies.
- Integration of all of UN agency responses so that prevention of workplace bullying/violence is addressed consistently in policies concerned with labour standards, OSH protection, human rights, the environment, trade and finance, corruption, corporate governance, social responsibility, and crime.
- Formulation of international best-practice guidelines for a whole-of-government, business and community approach to safeguarding within national jurisdictions.
- Establishment of reporting and auditing standards as part of an accreditation system for organizations and agencies that subscribe to international best practice in safeguarding.
- Scientific evaluation of pilot implementation processes involving integrated approaches by stakeholders across international and domestic spheres.

The current UN response to workplace bullying/violence proceeds explicitly through the ILO (2003a) *Code of Practice* against occupational violence, the WHO (1995), and implicitly through programs against inhumane treatment within a range of agency charters. Arguably, a more integrated international response could be achieved within a headline whole-of-UN agency global anti-workplace bullying/violence program. Gaining the commitment of UN agencies (e.g. United Nations Educational, Scientific and Cultural Organisation, World Trade Organisation, World Bank Group, International Monetary Fund, Food and Agriculture Organisation, and World Intellectual Property Organisation) to such an initiative would be important for leverage against the multiplicity of forms of workplace bullying/violence transmitted in the international arena.

The terms of a headline UN initiative could be *UN Action Against Workplace Bullying/Violence,* similar to the *UN Action Against Terrorism* (2004). The program could be initiated by a UN declaration that action to develop an international protocol would be pursued over the longer term. Such a declaration could be framed in terms that:

> Every human being in organizational operating environments is entitled to reasonable protection against inhumane acts that cause physical or psychological injuries, whether as employees, clients, contractors, suppliers, internees, or members of impacted communities.

The declaration could be accompanied by a statement of proposed terms for the development of an international protocol against workplace bullying/violence, as a basis for consultation.

Conclusion

Safeguarding against workplace bullying/violence has progressed from its roots in local/regional advocacy by diverse employee, professional, industry and regulatory interests to guidelines, and, in a few cases, to compliance standards. This progression of prevention has occurred mainly in wealthy industrialized western states. The fragmentation of responses and the lack of international best practice standards places the future of prevention within nation states at the whims of governments of the day. It leaves the international arena as a deregulated space in which workplace bullying/violence can flourish and be projected into nation states. The integration of policies across agencies concerned with labour standards, health, sustainability, justice, and security will be vital to realize benefits of safeguarding for all stakeholders. While the ILO (2003a) *Code of Practice* has initiated an international response, a whole-of-UN agency initiative set in train by a UN Declaration of an anti-workplace bullying/violence action program is now needed to accelerate safeguarding and further the integration of diverse responses. The networking of diverse global interests in advocacy for such a declaration is now called for.

Appendix 1
Crime Prevention in Small Business: Self-Audit Checklist

Claire Mayhew

This self-audit checklist is designed to assist small business owners and managers assess their vulnerability to a range of crimes, including violence.

This self-audit checklist is only a guide. Independent advice from an experienced security professional should be sought for specific sites and problems.

Armed Hold-Ups *When Open*

Risk Factors
Cash on site
Business open at night
Business on a main road
Business held-up in past 4 years
Only one worker on site
Staff willing to change large notes

Possible ways to reduce the risks				
Bright lighting inside so easily seen from street or mall	[] yes	[] no	[] average	[] n/a
All indoor lights working	[] yes	[] no	[] not always	
Windows have a clear view outside	[] yes	[] no	[] average	[] n/a
People in cash register area in clear view from street	[] yes	[] no	[] average	[] n/a
Time-delay access or drop safe near cash register	[] yes	[] no	[] not applicable	
Storage areas and staff rooms always locked	[] yes	[] no	[] not always	
Rear door always locked & bolted	[] yes	[] no	[] fire door	[] n/a
Wide counters between	[] yes	[] no	[] high counters	

226

Possible ways to reduce the risks – *continued*

Sign near door 'limited cash' on site	[] yes	[] no	[] not applicable
Height markers on exit doors, e.g. at 1.5, 1.75 & 2 metres	[] yes	[] no	
All staff have access to outside phone line	[] yes	[] no	[] limited
Emergency numbers listed on phone/phone programmed	[] yes	[] no	[] list near phone
Panic alarms at till	[] yes	[] no	
Tight cash control procedures	[] yes	[] no	[] average
Additional late-night/late afternoon cash procedures	[] yes	[] no	[] not applicable
Till float limit e.g. £50 or $150, with any more placed in safe	[] yes	[] no	[] not enforced
Silent alarm at main cash site linked with security firm	[] yes	[] no	
Staff wear personal alarms	[] yes	[] no	[] at high-risk times
Additional security measures if open after hours	[] yes	[] no	[] some [] n/a
Security courier service hired to deposit cash at bank	[] yes	[] no	[] usually
Cash taken to bank at random times by variable routes	[] yes	[] no	[] usually
Cash carried to bank in inconspicuous bag	[] yes	[] no	[] usually
Cash carried to bank by adult staff member	[] yes	[] no	[] usually
Strict opening and closing procedures	[] yes	[] no	[] if possible
'No working alone' policy	[] yes	[] no	[] not always
Back door watched whenever emptying rubbish	[] yes	[] no	[] bright light [] n/a
Staff told to do exactly what robbers say	[] yes	[] no	
Staff trained for hold-ups e.g. give robbers everything; *not* fight back; avoid sudden movements; avoid eye contact; remember description of robbers; *not* activate noisy alarms till robbers leave; lock premises immediately; and not talk to each other before police arrive	[] yes	[] no	[] a while ago

Shoplifting

> ### Risk factors
>
> Goods or materials on display to customers

Possible ways to reduce the risks

Price tags that cannot be switched between articles	[] yes [] no [] not applicable
Before sale, price ticket examined e.g. altered or torn	[] yes [] no [] if applicable
Non-removable wrapping on goods	[] yes [] no [] when possible
Electronic bar coding on all goods	[] yes [] no [] when possible
Cash register/stock management system with bar codes	[] yes [] no [] not applicable
Item/price ticketing records held at point-of-sale	[] yes [] no
Shelving is low to allow clear visibility across shop	[] yes [] no [] when possible
Displays arranged to minimize blind spots	[] yes [] no [] where possible
Convex mirrors in corners so shoplifters can't hide	[] yes [] no [] not applicable
Electronic sensor alarms emit a sound on entry/ exit	[] yes [] no [] quiet times only
Expensive goods in locked cabinets or behind counter	[] yes [] no [] when possible
Attached cables threaded through high-value items	[] yes [] no [] not applicable
All customers exit shop past well-lit staff area	[] yes [] no
Closed circuit television or cameras record all customers	[] yes [] no [] if affordable
Sign at entrance states 'video surveillance of shop'	[] yes [] no [] not applicable
Sign states all shoplifters prosecuted	[] yes [] no
Sign states condition of entry is bags may be inspected	[] yes [] no [] not permitted
Only one entry/exit open when few staff on site	[] yes [] no [] wheelchair access
Strict controls over goods taken to fitting rooms	[] yes [] no [] not applicable
'Hot products' for theft identified, watched & secured	[] yes [] no [] not applicable
All serial numbers of expensive goods recorded	[] yes [] no [] not applicable

Possible ways to reduce the risks – *continued*

Any waste wrapping or tickets found quickly reported	[] yes [] no [] not applicable
Large goods picked-up from secured yard	[] yes [] no [] not applicable
System to control drive-off without paying e.g. garage	[] yes [] no [] not applicable
Good customer service & attention at all times	[] yes [] no [] if not busy
Staff alert to 'browsing' customers	[] yes [] no
Staff watch all children & serve as quickly as possible	[] yes [] no [] when possible
Staff aware professional thieves may 'steal to order'	[] yes [] no
Staff alert to customers with large bags, strollers, coats	[] yes [] no
Staff aware thieves may be older, well-dressed, pregnant	[] yes [] no
Staff alert to gangs where 1 distracts and another steals	[] yes [] no [] may not be obvious
Staff know words to use when ask people to leave shop	[] yes [] no [] refer to manager
Staff trained on approaches to suspected shoplifters	[] yes [] no [] alert manager
Staff know what to do if shoplifter caught	[] yes [] no [] refer to manager
All shoplifters reported to police, even small item thefts	[] yes [] no [] adults only
Sales tallied daily and rolling stocktakes conducted	[] yes [] no [] weekly
Full inventory count e.g. when tax reports due	[] yes [] no [] more often

Break and Enter *When Closed*

> ### Risk Factors
>
> Business in high-crime area
> Business in isolated area
> Business in high vandalism area
> Cash, valuable goods, or 'hot products' kept on site when closed
> Cars stored on site overnight
> Places where offenders can hide inside before business closes
> e.g. bathroom or yard

Possible ways to reduce the risks

Internal bright lighting left on so easily seen from outside	[] yes	[] no	[] average
Bright lighting outside at night	[] yes	[] no	[] not yard
Motion-activated sensor lighting in yard areas	[] yes	[] no	[] not applicable
Alarms linked with security firm	[] yes	[] no	
Alarm activated by breaking window glass or impact sensor	[] yes	[] no	
24-hour security monitor at central site	[] direct line	[] mobile data back-up	[] no
Security guards patrol when closed	[] yes	[] no	[] occasional
All exterior doors are strengthened and have deadlocks	[] yes	[] no	[] partial
Exterior doors & windows do not have outside hinges	[] yes	[] no	
24-hour security cameras or closed circuit television	[] yes	[] no	[] not working
Security camera tape in locked place	[] yes	[] no	[] not applicable
Security grills on windows, skylights etc.	[] yes	[] no	[] some
Metal shutters fitted outside or inside	[] yes	[] no	[] roll-out mesh
Bollards fitted outside to prevent ram-raids (or large solid pot plant containers)	[] yes	[] no	[] not applicable
Store room doors secure from outside opening	[] yes	[] no	[] not applicable
External electric security fence installed	[] yes	[] no	[] not electric

Possible ways to reduce the risks – *continued*

Credit card machines etc secured in safe place after hours	[] yes	[] no	[] secure in spot	[] n/a
Empty cash register drawers left wide open overnight	[] yes	[] no	[] unlocked	
Expensive equipment secured	[] yes	[] no	[] not applicable	
Serial numbers of equipment recorded e.g. computer	[] yes	[] no	[] not applicable	
Indelible marking on expensive equipment & stock	[] yes	[] no	[] not applicable	
Cash and account records secured in separate places	[] yes	[] no	[] not applicable	
Computer system/software protected & secured	[] yes	[] no	[] not applicable	
Daily trading computer records backed-up & stored off-site	[] daily	[] no	[] not applicable	
Strict security over keys	[] yes	[] no	[] when possible	
Locks and keys, and routines, changed regularly	[] yes	[] no	[] not possible	

Credit Card, EFTPOS and Cheque Fraud

Warning signs and risks

Aggressive customers try to bully staff into acceptance
Customers appear unusually nervous
Customers take a long time to enter PIN numbers or to write signature
Customer unusually interested in shop credit limit
Customer details can be 'skimmed' off magnetic strip when card borrowed, lost or stolen
Insufficient money in account to cover cheque
Stolen credit cards used to purchase goods, particularly if offenders know PIN no

Possible ways to reduce the risks of credit card and EFTPOS fraud

Credit cards are examined for damage, imprinted mark, expiry & commencement dates & abnormal features	[] yes	[] no	[] usually
Number shown on EFTPOS machine compared to that embossed on card	[] always	[] no	[] not applicable
Docket signature carefully checked against that on card	[] always	[] no	
Amount, approved/declined & transaction type checked	[] yes	[] no	
Credit cards that cannot be read	[] refused	[] accepted with conditions	
Photographic identification required if any irregularities	[] yes	[] no	[] sale refused
Credit centre clearance for all high value goods	[] yes	[] no	[] not applicable
Card returned to customer only after sale finalized	[] yes	[] no	[] sometimes
Personal carbon copies of record always secured	[] yes	[] no	[] destroyed
If sale blocked, credit centre directions always followed	[] yes	[] no	
Warning bulletins/lists checked daily	[] yes	[] no	[] not sent to us
EFTPOS terminals never left unattended	[] yes	[] no	[] secured
Staff well trained/re-trained in EFTPOS procedures	[] yes	[] no	[] usually
Ensure customers can touch EFTPOS machine	[] only to enter PIN no.		

Possible ways to reduce the risks of cheque fraud

Strict cheque acceptance policies enforced	[] yes	[] no	[] usually
All cheques	[] accepted if equifax cleared		
Reverse side endorsed with person's full name, address, contact numbers and details of ID presented	[] yes	[] no	[] + staff name
'Cash' cheques	[] refused	[] accepted if know person	
Cheques presented just before holiday or weekend	[] refused	[] accepted if know person	
Cheques above item purchase price	[] refused	[] accepted if small change	
Cheque signature compared against identification	[] always	[] photo ID	
Two forms of photographic identification required	[] yes	[] no	[] only one ID
Staff trained in how to check photo identification	[] yes	[] no	[] sometimes
Cheques signed in front of staff	[] always	[] usually	[] no
Post-dated cheques	[] refused	[] accepted if know person	
Cheques from out-of-town people	[] refused	[] accepted if Equifax	[] accepted with ID [] n/a
Records and/or photocopies of details on cheques kept	[] yes	[] no	
Bank phoned to confirm clearance of bank checks	[] yes	[] no	[] not applicable
Bank cheques presented after bank closed for the day	[] refused	[] accept	[] depends [] n/a
A limit is set on amount payable by cheque e.g. $200 or £50	[] yes	[] no	[] usually [] n/a
Cheques are always crossed	[] yes	[] no	
Travelers cheques countersigned, passport/bank verified	[] yes	[] no	[] not applicable
All cheques deposited and cleared asap	[] yes	[] no	

Possible counterfeit currency

Watermarks & opaque window emblems checked	[] always	[] big notes	[] when warned

Refund Fraud

Risks

Aggressive customers who try to force a questionable refund
Stolen goods presented for cash refund
Stolen goods presented for refund using discarded receipts
Sale goods submitted for full-price refund
Goods bought or stolen from another shop submitted for refund
Goods paid for with stolen credit card presented for refund
Goods paid for with stolen cheque presented for refund
Goods paid for with stolen gift voucher presented for refund
Ticket-switched goods presented for refund at higher price
Stock donated to charity presented for a refund

Possible ways to reduce the risks of refund fraud

Receipt required for all refunds	[] yes	[] no	[] unless faulty
Minimum requirements set for refund e.g. faulty goods	[] yes	[] no	[] exchange
'Receipt required for refund' printed on all price tickets	[] yes	[] no	
Refunds limited to exchange of goods only (if no receipt)	[] yes	[] no	[] not allowed
Price tickets during sales are different to full-price items	[] yes	[] no	
Refund receipts marked or kept to prevent re-use	[] yes	[] no	
Refund receipts are different colour and shape to others	[] yes	[] no	
All price tickets have shop name printed on them	[] yes	[] no	[] sometimes
Computer check of items to be refunded	[] yes	[] no	[] not available
Refund of expensive items worn once e.g. evening dress	[] refused	[] depends	[] refunded
Two forms of photographic identification required	[] yes	[] 1 only	[] if no receipt
All refunds double-checked by supervisor	[] yes	[] no	[] if possible
All refunds require authorization by two people	[] yes	[] no	[] if possible
Names & addresses of refunds collated	[] yes	[] no	[] sometimes
Refunds over £30/$100 paid by cheque to account holder	[] yes	[] no	[] not allowed
Refunds for goods paid by credit card returned onto card	[] yes	[] no	
Goods purchased by cheque not refunded unless cleared	[] yes	[] no	[] if equifax

Possible ways to reduce the risks of refund fraud – *continued*

Bar code system means refund prices can't change	[] yes	[] no	[] not applicable
Sign inside shop states business refund policy	[] yes	[] no	
Closed circuit television (or camera) faces sale counter	[] yes	[] no	
Camera linked with electronic cash register	[] yes	[] no	[] not affordable
Layout of sale/refund counter stops 'pick-up' fraud	[] yes	[] no	[] not applicable
Care taken with disposal of damaged stock	[] yes	[] no	
Staff well trained in refund policies & procedures	[] yes	[] no	[] need more
Price reductions carefully recorded in case of refunds	[] yes	[] no	[] too time-costly
Cash refund items checked against sales and inventory	[] yes	[] no	[] not applicable
Intermittent mail survey of customers who return items	[] yes	[] no	[] not applicable
All cheque & card sales have details placed on database	[] yes	[] no	
All refunds recorded in accounting format database	[] yes	[] no	
Fraudulent cheque/refund details circulated	[] yes	[] no	

Scams and Other Frauds

Risks
Supply of fewer goods than purchased
Supply of poorer quality goods than purchased
Arrival of an invoice for non-purchased and unknown goods
Phone call from telemarketer using 'hard sell'
Unauthorized access to business computer
Unauthorized access to company files
Business deals with an insolvent/near bankrupt company
Employees or managers spending without authorization

Possible ways to reduce the risks

Check & balance procedures in invoice/statement processing	[] yes	[] no	[] basic
Invoices *always* checked against order numbers	[] yes	[] no	
Comprehensive order number system in place	[] yes	[] no	
Suspect & ambiguous invoices	[] not paid	[] questioned	
Payment demands for non-ordered goods (e.g. by phone)	[] refused	[] checked	
Reply letters to suspect businesses	[] not signed by financial account holder signature		
Careful checks on businesses ordering goods on credit	[] all new accounts	[] at regular intervals	
Business credit limit checked before goods issued	[] yes	[] no	[] not applicable
Business credit approval procedures make fraud difficult	[] yes	[] no	[] don't give credit
Guidelines to prevent fraud developed & used	[] yes	[] no	[] very basic
Flow chart of financial procedures and controls	[] yes	[] no	[] not applicable
Most common types of fraud identified	[] yes	[] no	
Assets, liabilities & accounts vulnerable to fraud noted	[] yes	[] no	
Business uses a unique 'okay to pay' stamp on invoices	[] yes	[] no	[] 'signed off'
Internal fraud controls established	[] yes	[] no	
Fraud vulnerability audits conducted	[] yes	[] no	[] if necessary

Possible ways to reduce the risks – *continued*

Business credit card number	[] never disclosed to unauthorized people			
Credit card belonging to *your business* always secured	[] yes	[] no	[] usually	[] n/a
Business credit card never used in suspect businesses	[] yes	[] no	[] not applicable	
Can gift vouchers be redeemed for cash	[] yes	[] no	[] not applicable	
Gift vouchers are made out to named person	[] yes	[] no	[] not applicable	
All gift vouchers have unique number	[] yes	[] no	[] not applicable	
Gift voucher record of sales kept	[] yes	[] no	[] not applicable	
Unused gift vouchers treated as if they were cash	[] yes	[] no	[] not applicable	
Staff signature on gift voucher always double-checked	[] yes	[] no	[] not applicable	
All trade referees checked e.g. electrician	[] yes	[] no	[] new people	
The national Securities and Investment Commission public register checked for all large contracts	[] yes	[] no	[] not applicable	

Delivery Theft and Fraud

Risk Factors

Deliveries made out of regular work hours
Delivery people left alone in yard or storeroom
Stock left unattended in delivery area
Rubbish bins in delivery area
Procedures that allow short loads to arrive undetected
Procedures that allow short loads to leave undetected
Procedures that allow damaged goods to go undetected
Procedures that allow the non-arrival of posted goods to go undetected
Procedures that allow employees to collude with delivery drivers

Possible ways to reduce the risks

Orders and invoices carefully checked before signature	[] yes	[] no	[] if possible
Number of cartons checked against delivery docket	[] yes	[] no	
Deliveries without invoices	[] refused	[] docket endorsed	[] accepted
Damaged goods	[] refused	[] docket endorsed	[] depends
Opened cartons or containers with broken seals	[] refused	[] docket endorsed	
Short loads	[] refused	[] docket endorsed	
Goods left unattended during loading/unloading	[] yes	[] never	[] rarely
Cartons in delivery truck (e.g. for other customers)	[] checked	[] ignored	[] can't check [] n/a
Spot check unloading of delivery vans leaving site	[] yes	[] no	[] can't force [] n/a
Rubbish bins in yard spot-checked daily	[] yes	[] no	[] occasional [] n/a
Empty cartons stored in separate area	[] yes	[] no	
Staff park vehicles out of delivery yard	[] yes	[] no	
Manager checks for collusion between staff & drivers	[] yes	[] no	[] if warranted
Secure lock up of delivery area & yard overnight	[] yes	[] no	[] not applicable
Stockroom counts recorded and checked	[] daily	[] weekly	[] part of overall
Tally of inventory, deliveries & sales recorded	[] daily	[] weekly	[] 3 months [] n/a
Double-check of customer receipt of goods	[] on the spot	[] weekly	[] occasional [] n/a

Employee Theft or Fraud

Risks

Cash register void function used to pocket customer cash
Employees serve their close friends and relatives
Friends slip by with unpaid articles, or buy at 'special' discounts not authorized by business
Customer receipts withheld & refunds claimed later
Employees damage goods & then buy at discount
Employees submit false refund claims
Employee sends stock to their home

Possible ways to reduce the risks

Strict separation of cash handling and accounting roles	[] yes	[] no	
Security protection of computer file and account records	[] yes	[] no	
Regular checking of inventory against account records	[] yes	[] no	[] when possible
Constant checks on balance/change at point-of-sale	[] yes	[] no	[] end of day
All manual entries on point-of-sale record picked up	[] yes	[] no	[] not applicable
Surprise cash counts at point-of-sale	[] yes	[] no	[] if warranted
Random checks on employee scanning at point-of-sale	[] yes	[] no	
Random checks on refunds at point-of-sale	[] yes	[] no	
Copy of refund slip given to buyer	[] yes	[] hand marked if can't print out	
All refund slips destroyed or marked immediately	[] yes	[] no	[] sent to accounts
Voids at point-of-sale collated and tracked daily	[] yes	[] no	[] weekly
Record of price 'over-rides' collated/printed out	[] daily	[] no	[] weekly
Shortfalls from common-use tills tracked daily	[] yes	[] no	[] weekly
Random spot checks on high-value goods during day	[] yes	[] no	
Closed circuit television records cash register activities	[] yes	[] no	

Possible ways to reduce the risks – *continued*

Background checks before hire of new staff	[] yes	[] no	
Detailed orientation for new staff	[] yes	[] no	[] on-the-job
In-depth staff training/re-training on procedures	[] yes	[] no	[] occasional
Employees paid via commission have strict guidelines	[] yes	[] no	[] not applicable
All staff discounts signed for & authorized by 1 person	[] yes	[] no	
All staff told of penalties for dishonesty	[] yes	[] no	[] in orientation
Prohibition of unauthorized discounting/mark downs	[] yes	[] no	
Employees told cheap deals for friends prohibited	[] yes	[] no	
All personal staff belongings stored in secure room	[] yes	[] no	
Random search of bags of *all* staff at end of day	[] yes	[] no	[] if union agrees [] n/a
Background checks on contractors visiting site	[] yes	[] no	[] sometimes
Employees who defraud are dismissed	[] yes	[] no	[] warned 3 times
Key access restricted, with no labels attached to keys	[] yes	[] no	
Strict security control over keys to account files	[] yes	[] no	[] not applicable
Security control over mark down stickers and tags	[] yes	[] no	[] not applicable
Close supervision and security over safe	[] yes	[] no	[] not applicable
Security maintained over stockrooms	[] yes	[] no	
Unused cheques in business chequebooks counted daily	[] yes	[] no	[] weekly [] n/a
Daily count of cash box in office by two people	[] yes	[] no	[] weekly
'Mystery' customers check service	[] yes	[] no	[] occasional
Daily owner/manager 'walk-through'	[] yes	[] no	[] weekly

Bibliography

AccountAbility, *Standard 1000*. See: www.accountability.org.uk

Agamben, G. 'Security and Terror', translated by Emcke, C. *Theory and Event,* 5(4) (2002). See: http:muse.jhu.edu/journals/theory_and_event/v005/5.4agamben.html

Ahmed, E. Harris, N. Braithwaite, J. and Braithwaite, V. *Shame Management through Reintegration* (UK: Cambridge University Press, 2001).

Alexander, B. Franklin, G. and Wolf, M. 'The sexual assault of women at work in Washington State, 1980–89', *American Journal of Public Health,* 84(4) (1994) 640–42.

Ali, T. *The Clash of Fundamentalism: Crusades, Jihads and Modernity* (London: Verso, 2002).

American Psychiatric Association (APA), *Diagnostic and Statistical Manual of Mental Disorders 1V – Text Revision* (Washington DC: American Psychiatric Association, 2000).

Amnesty International, 'A human rights scandal: Violence against women' (Berne, Switzerland: Speech delivered by Irene Kahn, Secretary General Amnesty International for the International Day for the Elimination of Violence against Women, 25 November, 2003). http://web.amnesty.org/library/print/ENGACT770182003.

Amnesty International, 'The pain merchants: Security equipment and its use in torture and other ill treatment' (2003) 1–17. See: http://web.amnesty.org/library/print/ENG400082003

Ananthaswamy, A. 'The way of the gun', *New Scientist* (12 July 2003) 8–10.

Anderson, B. *Imagined Communities: Reflections on the Origin and Spread of Nationalism* (London: Verso, 1983).

Andersson, L. and Pearson, C. 'Tit for tat? The spiralling effect of incivility in the workplace', *Academy of Management Review,* 24(3) (1999) 452–71.

Anon, 'The rise of corporate propaganda', *New Internationalist* (July 1999) 18–9.

Ash, T. *History of the Present* (London: Allen Lane, 1999).

Ashforth, B. and Humphrey, R. 'Emotion in the workplace', *Human Relations,* 48 (1995) 97–125.

Australian Broadcasting Commission (ABC) 'Do Australia's defamation laws stifle freedom of expression?' *Radio National Law Report* with Damien Carrick (ABC, 6 March, 2001). http://www.abc.net.au/rn/talks/8.30/lawrpt/stories/ s255552.htm

Australian Broadcasting Commission (ABC), 'Background briefing' (Canberra: Radio National, 23 May 2000).

Australian Institute of Criminology and The Division of Worksafe Health and Safety (Qld), *Occupational Violence: Were you Threatened at Work Today?* (Brisbane, Bardon Professional Development Centre, Occupational Violence Seminar, 10–12 February 1993).

Axelby, C. *Risky Business: A Useful Publication for Employers for Preventing and Resolving Workplace Bullying* (Brisbane: Queensland Working Women's service, 2000)

Babiak, P. 'When psychopaths go to work: a case study of an industrial psychopath', *Applied Psychology; An International Review*, 44 (2) (1999) 171–88.

Banks, M. Clegg, C. Jackson, P. Kemp, E. Stafford, M. and Wall, T. 'The use of the General Health Questionnaire as an indicator of mental health in occupational studies', *Journal of Occupational Psychology*, 53 (1980) 187–94.

Barclay, G. Tavares, C. Kenny, S. Siddique, A. and Wilby, E. *International Comparisons of Criminal Justice Statistics 2001* (London: Development and Statistics Directorate, Home Office Research, The Home Office, 2003).

Barclay, G. and Tavares, C. *International Comparisons of Criminal Justice Statistics* (London: Development and Statistics Directorate, Home Office Research, The Home Office, 2002).

Bareja, M. and Charlton, K. 'Statistics on juvenile detention in Australia: 1981–2002', *Technical and Background Paper Series No. 5* (Canberra: Australian Institute of Criminology, 2003).

Barker, J. *The No-Nonsense Guide to Terrorism* (Oxford: New Internationalist publications in association with Verso, 2003).

Barker, M. and Sheehan, M. 'Bully or advocate? Reflections on the lawyer client relationship', in Sheehan, M. Ramsay, S. and Patrick, J. (eds) *Transcending Boundaries: Integrating People, Processes and Systems* (Brisbane: Griffith University School of Management, Conference Proceedings 6–8 September 2000).

Barling, J. 'The prediction, experience, and consequences of workplace violence', in Van den Bos, G. and Bulatao, E. (eds) *Violence on the Job* (Washington DC: American Psychological Association, 1996).

Baudrillard, J. 'L'esprit du terrorisme', *Le Monde*, 3 November 2001.

Bauman, Z. *Society under Seige* (UK: Polity, 2002).

Bauman, Z. *Postmodern Ethics* (Oxford: Blackwell, 1993).

Bauman, Z. *Modernity and the Holocaust.* (Oxford: Blackwell, 1991).

Beck, A. and Willis, A. *Crime and Security: Managing the Risk to Safe Shopping* (Leicester: Perpetuity Press, 1995).

Beck, A. Gill, M. and Willis, A. 'Violence in retailing: physical and verbal victimisation of staff', in Gill, M. (ed.) *Crime at Work: Studies in Security and Crime Prevention* (Leicester: Perpetuity Press, 1994).

Beck, U. 'From industrial society to risk society: Questions of survival, structure and ecological enlightenment', *Theory, Culture and Society*, 9 (1992) 97–123.

Beed, C. *Cultures of Secrecy and Abuse: A Paradox for Churches* (Hawthorn: Beed, 1998).

Bellamy, L 'Situational crime prevention and convenience store robbery', *Security Journal*, 7 (1996) 41–52.

Berman, M. *All That is Solid Melts* (London: Verso, 1982).

Bernstein, R. *The New Constellation* (UK, Cambridge: Polity Press, 1991).

Bibby, P. *Personal Safety for Health Care Workers* (UK: report commissioned by the Suzy Lamplugh Trust, 1995).

Biddle, E. and Hartley, D. 'The cost of workplace homicides in the USA, 1990–1997' (Montreal: 6th World Conference on Injury Prevention and Control, Les Presses de Montreal, May 2002) 421–22.

Björkqvist, K. Österman, K. and Hjelt-Bäck, M. *Aggression Among University Employees*, *Aggressive Behaviour*, (20) (1994) 173–84.

Bordieu, P. *Distinction* (London: Routledge, 1984).

Bowie, V. 'Defining violence at work: a new typology', in Gill, M. Fisher, B. and Bowie, V. (eds) *Violence at Work: Causes, Patterns and Prevention* (UK: Willan publishing, 2002) 1–20.

Boyd, C. 'Customer violence and employee health and safety', *Work, Environment and Society,* 16(1) (2002) 151–169.

Braude, J. *The New Iraq* (Sydney: Harper Collins Publishers, 2003).

Briscoe, S. and Donnelly, N. 'Assaults on licensed premises in inner-urban areas', *Alcohol Studies Bulletin,* 2 (2001a) 1–16.

Briscoe, S. and Donnelly, N. 'Temporal and regional aspects of alcohol-related violence and disorder', *Alcohol Studies Bulletin,* 1 (2001b) 1–15.

Brodsky, C. *The Harassed Worker* (Lexington MA: DC Heath and Company, 1976).

Brookes, J. and Dunn, R. 'The incidence, severity and nature of violent incidents in the emergency department', *Emergency Medicine.* 9 (1997) 5–9.

Budd, T. *Violence at work: New Findings from the 2000 British Crime Survey* (UK: Health and Safety Executive and the Home Office, July 2001).

Budd, T. *Violence at work: Findings from the British Crime Survey* (London: HMSO, the Home Office and the Health and Safety Executive, 1999).

Bureau of Crime Statistics & Research (BCS&R) *Top 25 Local Government Areas (LGAs) 2001 (Assault)* (Sydney: NSW Bureau of Crime Statistics and Research, 2003a). See: www.lawlink.nsw.gov.au

Bureau of Crime Statistics & Research (BCS&R) *Top 25 LGAs 2001 (Sexual Assault)* (Sydney: NSW Bureau of Crime Statistics and Research, 2003b).

Bureau of Crime Statistics & Research (BCS&R) *Who are the Offenders?* (Sydney: NSW Bureau of Crime Statistics and Research, 2003c).

Bureau of Crime Statistics and Research (BCS&R), *New South Wales Criminal Court Statistics 1999* (Sydney: NSW Bureau of Crime Statistics and Research, 2000).

Caldeira, T. 'Fortified enclaves: The new urban segregation', *Public Culture,* 8(2) (1996) 303–29.

Californian Division of Occupational Safety and Health Administration (CAL/OSHA), *Guidelines for Security and Safety of Health Care and Community Service Workers* (San Francisco: Division of Occupational Safety and Health, 1998).

Caple, D. 'Reduction of occupational violence through architectural controls and work procedures', *Journal of Occupational Health and Safety – Australia and New Zealand,* 16(5) (2000) 437–43.

Capozzoli, T. and McVey, R. *Managing Violence in the Workplace* (Florida, Delray Beach: St. Lucien Press, 1996).

Cascio, W. (1993) 'Downsizing: What do we know? What have we learned?' *Academy of Management Executive,* 7(1) 95–104. www.mgmt.utoronto.ca/ ~evans/teach363/downsize/respes.htm

Casilli, A. *Stop Mobbing* (Roma: Derive Approdi, 2000).

Castillo, D. and Jenkins, L. 'Industries and occupations at high risk for work-related homicide', *Journal of Occupational Medicine,* 36(2) (1994) 125–32.

Catley, B. and Jones, C. 'On the undecidability of violence' (Dublin: Trinity College, paper presented at the Standing Conference on Organisational Symbolism (SCOS), 30th June to 4th July 2001).

Chapman, S. 'The use of the general health questionnaire in the Australian defence force: a flawed but irreplaceable measure?' *Australian Psychologist*, 36(3) (2001) 244–49.

Chappell, D. 'Violence and its victims in the transportation industry workplace', in Friday, P. and Kirchhoff, G. (eds) *Victimology at the Transition From the 20th to the 21st Century*, (Monchengladbach, Federal Republic of Germany: Shaker Verlag and World Society of Victimology publishing, 2000) 289–312.

Chappell, D. and Di Martino, V. *Violence at Work*, 2nd edition (Geneva: International Labour Office, 2000).

Chengappa, A. 'Hindu-Muslim divide', *India Today* (New Delhi: Living India, January 31st 1993a).

Chengappa, A. 'Rocket technology: Joining the big boys', *India Today* (New Delhi: Living India, January 31 1993b).

Chengappa, R. and Menon, R. 'The new battlefields', *India Today* (New Delhi: Living India, January 31st 1993c).

Cherry, D. and Upston, B. *Managing Violent and Potentially Violent Situations: A Guide for Workers and Organisations*, (Victoria: Centre for Social Health, Heidelberg West, 1997).

Chomsky, N. *September 11* (Australia: Allen and Unwin, 2001).

Chua, A. *World on Fire: How Exporting Free Market Democracy Breeds Ethnic Hatred and Global Instability* (New York: Random House, 2003).

Clarke, R. (ed.) *Situational Crime Prevention: Successful Case Studies* (New York: Harrow and Heston, 1992).

Clayton, A. Mayhew, C. and Quinlan, M. 'Injury, illness, deaths, suicide, substance abuse, and medical care availability' (Melbourne: Savoy Park Plaza Hotel, paper presented to the *Seafarers' Welfare Forum*, 22–23 August, 2000).

Clegg, S. *Modern Organizations: Organization Studies in a Postmodern World* (London: Sage, 1990).

Cohen, S. and Grace, D. 'Ethics and sustainability', in CCH and Freeholds *Australian Master OHS and environment guide 2003* (Australia: CCH, 2002) 443–62.

Cole, L. Grubb, P. Sauter, S. Swanson, N. and Lawless, P. 'Psychosocial correlates of harassment, threats and fear of violence', *Scandinavian Journal of Work, Environment and Health*, 23(6) (1997) 450–57.

Community Aid Abroad, *The Nike Watch Campaign* (2000). See: www.caa.org.au/campaigns/nike

Connolly, C. 'The role of private security in combating terrorism', *Journal of Homeland Security*, 30 September 2003). See: http://www.homelandsecurity.org/journal/Commentary/displayCommentary.asp?commen

Conroy, J. *Unspeakable Acts: Ordinary People* (New York: Knopf, 2000).

Copjec, J. 'Evil in the time of the finite world', in Copjec, J. (ed.) *Radical Evil* (London: Verso, 1996a).

Copjec, J. *Radical Evil* (London: Verso, 1996b).

Couper, A. 'Discussion comments', in *Proceedings of a Research Workshop on Fatigue in the Maritime Industry* (Cardiff: Seafarers International Research Centre, University of Wales, April 23–25, and May 7–9 1996).

Coyne, L. Seigne, E. and Randall, P. 'Predicting workplace victim status from personality', *European Journal of Work and Organizational Psychology*, 9 (2000) 335–49.

Crowe, T. and Adams, J. 'CPTED: applying principles to offices and large sites', *Security Australia,* 15 (2) (1995) 36–9.

Cunneen, C. Fraser, D. and Tomsen, S. (eds) *Faces of Hate: Hate Crime in Australia* (Annandale: Hawkins Press, 1997).

Daniel, A. *Scapegoats for a Profession: Uncovering Procedural Injustice* (Sydney: Harwood Academic Publishers, 1998).

Debray, R. *Critique of Political Reason* (London: New Left Press, 1983).

De Croon, E. Blonk, R. de Zwart, B. Frings-Dresen, M. and Broersen, J. 'Job stress, fatigue, and job dissatisfaction in Dutch lorry drivers: towards an occupation specific model of job demands and control', *Occupational and Environmental Medicine,* 59(6) (2002) 356–61.

DeGeneste, H. and Sullivan, J. *Policing Transportation Facilities* (Illinois: Charles C. Thomas, 1994).

Deleuze, G., and Guattari, F. *A Thousand Plateaus* (Minnesota: University of Minnesota Press, 1987).

De Maria, W. *Deadly Disclosures: Whistleblowing and the Ethical Meltdown of Australia.* (Kent Town, South Australia: Wakefield press, 1999).

Department of Industrial Relations (DIR), Queensland, *Prevention of Workplace Harassment Advisory Standard* (Brisbane: Department of Industrial Relations, 2004).

Derrida, J. *Positions* (Chicago: University of Chicago Press, A. Bass Trans, 1981).

Dietz, J. Robinson, S. Folger, R. Baron, R. and Schulz, M. 'The impact of community violence and an organization's procedural justice climate on workplace aggression', *The Academy of Management Journal,* 46(3) (2003) 317–26.

Di Martino, V. *Workplace Violence in the Health Sector Country Case Studies, Synthesis Report.* (Geneva: International Labour Office, International Council of Nurses, World Health Organization and Public Services International, Joint Programme on Workplace Violence in the Health Sector, 2002). See: www.ilo/public/ english/dialogue/sector/papers/health/violence-ccs.pdf or www.icn.ch

Di Martino, V. 'Violence at the workplace: the global challenge' (Johannesburg: International Conference on Work Trauma, November 8–9, 2000).

Di Martino, V. Hoel, H. and Cooper, C. *Preventing Violence and Harassment in the Workplace.* (Dublin: European Foundation for the Improvement of Living and Working Conditions, 2003.) See: www.eurofound.eu.int

Division of Workplace Health and Safety (DWH&S), *Workplace Bullying: an Employer's Guide* (Brisbane: Division of Workplace Health and Safety, 1998).

Division of Workplace Health and Safety, *Workplace Health and Safety Act 1995* (Brisbane: Queensland Government, 1995).

DSM III-R, *Diagnostic and Statistical Manual of Mental Disorders* (Washington DC: American Psychiatric Association, 1987).

Dow Jones Sustainability Index.

Dunlap, A. with Andelman, B. *Mean Business: How I Save Bad Companies And Make Good Companies Great* (New York: Times Business, Random House, 1996).

Easton, D. 'Drivers as workers, vehicles as workplaces', in STAYSAFE 36, *Drivers as Workers, Vehicles as Workplaces: Issues in Fleet Management* (Sydney: Parliament House, ninth report of the Joint Standing Committee on Road Safety of the 51st Parliament, edited transcripts of a seminar, 29 April 1997) 157–61.

Ehrenreich, B. *Nickel and Dimed: On (Not) Getting By in America* (New York: Metropolitan Books, 2001)

Ehrenreich, B. *Fear of Falling: The Inner Life of the Middle Class* (New York: Harper Perennial, 1990).

Einarsen, S. 'Harassment and bullying at work. A review of the Scandinavian approach', *Aggression and Violent Behaviour*, 5(4) (2000a) 371–401.

Einarsen, S. 'Bullying and harassment at work: Unveiling an organisational taboo' in Sheehan, M. Ramsay, C. and J. Patrick, J. (eds) *Transcending Boundaries* (Brisbane: Conference Proceedings, September 2000b)

Einarsen, S. Hoel, H. Zapf, D. and Cooper, C. (eds) *Bullying and Emotional Abuse in the Workplace* (London: Taylor Francis, 2003).

Einarsen, S.H. Hoel, D. Zapf and C. Cooper 'The concept of bullying at work: the European tradition', in S. Einarsen, H. Hoel, D. Zapf and C. Cooper (eds) *Bullying and Emotional Abuse in the Workplace: International Perspectives in Research and Practice* (London: Taylor & Francis, 2003b).

Einarsen, S. and Mikkelsen, E. 'Individual effects of exposure to bullying at work', in Einarsen, S. and Hoel, H. Zapf, D. and Cooper, C. (eds) *Bullying and Emotional Abuse in the Workplace: International Perspectives in Research and Practice* (UK: Taylor Francis, 2003) 127–44.

Einarsen, S. Matthiesen, S. and Mikkelsen, E. *Tiden Leger alle sår? Senvirkninger af Mobbing I Arbeidslivet* (Does time heal all wounds? Long term effects of exposure to bullying at work) (Bergen: University of Bergen, 1999).

Einarsen, S. Raknes, B. and Matthiesen, S. 'Bullying and harassment at work and their relationships to work environment quality: An exploratory study', *The European Work and Organizational Psychologist*, 4 (1994a) 381–401.

Einarsen, S. Raknes, B. Matthiesen, S. and Hellesoy, O. *Mobbing og Harde Personkonflikter. Helsefarlig Samspill pa[o] Arbeidsplassen* (bullying and harsh interpersonal conflicts) (Bergen: Sigma Forlag, 1994b).

Elston, M. Gabe, J. Denney, D. Lee, R. and O'Beirne, M. 'Violence against doctors: A medical(ised) problem? The case of National Health Service general practitioners', *Sociology of Health and Illness*, 24(5) (2002) 575–98.

Elzinga, A. 'Security of taxi drivers in the Netherlands: Fear of crime, actual victimization and recommended security measures', *Security Journal*, 7(3) (1996) 205–10.

Epstein, S. 'The implications of cognitive experiential self-theory for research in social psychology and personality', *Journal for the Theory of Social Behaviour*, 15(3) (1985) 282–310.

European Parliament (EP), 'Resolution on harassment at the workplace 2001/2339 (INI)', *Official Journal of the European Communities*, 20 September, 2001.

Farrell, G. 'Aggression in clinical settings: nurses' views – a follow-up study', *Journal of Advanced Nursing*, 29 (3) (1999) 532–41.

Featherstone, M. 'In pursuit of the postmodern: An introduction', *Theory, Culture and Society*, 5(2–3) (1998) 195–218.

Fisher, J. Bradshaw, J. Currie, B. Klotz, J. Robins, P. Searl, K. and Smith, J. *Context of Silence: Violence and the Remote Area Nurse.* (Rockhampton: Central Queensland University, Faculty of Health Science, 1995).

Fisher, B. and Gunnison, E. 'Violence in the workplace: gender similarities and differences', *Journal of Criminal Justice*, 29 (2001) 145–55.

Fisher, B. Jenkins, E. and Williams, N. 'The extent and nature of homicide and non-fatal workplace violence in the United States: Implications for prevention and security', in Gill, M. (ed.) *Crime At Work: Increasing the Risk for Offenders* (Leicester: Perpetuity, 1998).

Fisher, B. and Looye, J. 'Crime and small businesses in the Midwest: an examination of overlooked issues in the United States', *Security Journal*, 13 (1) (2000) 45–72.

Flannery, R. 'Violence in the workplace, 1970–1995: A review of the literature', *Aggression and Violent Behavior*, 1(1) (1996) 57–68.

Foucault, M. 'Governmentality', in Burchell, G. Gordon, C. and Miller, P. (eds) *The Foucault Effect* (London: Harvester Wheatsheaf, 1991) 87–105.

Foucault, M. *The History of Sexuality* (Ringwood, Victoria: Penguin, trans. R. Hurley, 1981).

Foucault, M. *Discipline and Punish: the Birth of the Prison* (New York: Pantheon, trans. Sheridan, A. 1977).

Foucault, M. *Surveiller et Punir* (Discipline and Punishment) (Paris: Editions Gallimard, 1975).

Foucault, M. *The Archaeology of Knowledge* (London: Tavistock, 1972).

Fukuyama, F. *The End of History and the Last Man* (New York: Free Press, 1992).

Galea, S. Ahern, J. Resnick, H. Kilpatrick, D. Bucuvalas, M. Gold, J. and Vlahov, D. 'Psychological sequelae of the September 11 terrorist attacks in New York city', *New England Journal of Medicine*, 346(13) (2002) 982–87.

Gaymer, J. 'Assault course', Occupational *Health*, 51(2) (1999) 12–13.

Girard, R. *Violence and the Sacred* (London: Johns Hopkins University press, trans. Gregory, P. 1977).

Glasl, F. *Konfliktmanagement. Ein Handbuch für Führunskräafte und Berater* (Conflict management: A handbook for managers and consultants), 4th edition (Bern, Switzerland: Haupt, 1994).

Glassner, B. *The Culture of Fear: Why Americans are Afraid of the Wrong Things* (New York: Basic Books, 1999).

Global Reporting Initiative, *Sustainability Reporting Guidelines* (Boston: Global Reporting Initiative, 2002). See: www.globalreporting.org

Goldberg, D. *The Detection of Psychiatric Illness by Questionnaire* (London: Oxford University Press, 1972).

Goldberg, D. Gater, R. Sartorius, N. Ustun, T. Piccinelli, M. Gureje, O. and Rutter, C. 'The validity of two versions of the GHQ in the WHO study of mental illness in general health care', *Psychological Medicine*, 27 (1997) 191–97.

Goldberg, D. and Williams, P. *A User's Guide to the General Health Questionnaire.* (Windsor, UK: NFER-Nelson, 1988, reprinted 1991).

Goodchild, M. and Duncan-Jones, P. 'Chronicity and the general health questionnaire', *British Journal of Psychiatry*, 146 (1985) 55–61.

Gordon, D. and Risley, D. *The Costs to Britain of Workplace Accidents and Work-related Ill-Health in 1995/6*, second edition (London: HSE Books, 1999).

Gottfredson, D. 'School-based crime prevention', in Sherman, L. Gottfredson, D. MacKenzie, D. Eck, J. Reuter, P. and Bushway, S. *Preventing Crime: What Works, What Doesn't, What's Promising* (Washington DC: National Institute of Justice, 1998). See: www.ncjrs.org/works/chapter5.htm

Grabosky, P. and Duffield, G. 'Red flags of fraud', *Trends and Issues in Crime and Criminal Justice*, number 200, (Canberra: Australian Institute of Criminology, 2001).

Graetz, B. 'Multidimensional properties of the general health questionnaire', *Social Psychiatry and Psychiatric Epidemiology*, 26 (1991) 132–38.

Graycar, A. 'Violence in the workplace: personal and political issues', paper presented at the *Security in Government Conference* (Canberra: 30 April to 2 May 2003).

Graycar, A. 'Small business against crime: situational strategies in action', paper presented at the *CPTED Conference* (Brisbane: 24 September 2001).

Grossman, L. 'The reputational panorama', in CCH and Freeholds *Australian Master OHS and environment guide 2003* (Australia: CCH, 2002) 427–43.

Guattari, F. *Molecular Revolution* (Aylesbury: Penguin, 1984).

Hage, G. *Against Paranoid Nationalism* (Annandale: Pluto, 2003).

Haines, F. 'Towards understanding globalisation and control of corporate harm: A preliminary criminological analysis', *Current Issues in Criminal Justice,* 12(2) (2000) 166–80.

Haines, F. *Taxi driver Survey – Victoria: Understanding Victorian Taxi Drivers' Experiences of Victimisation and Their Preferred Preventative Measures,* (Melbourne: University of Melbourne, Criminology department, report for the Victorian Taxi Driver Safety Committee, 1997).

Hannett, J. 'Voices from the front line', paper presented at the Health and Safety Executive *Tackling Work-related Violence – Putting Policies into Practice* conference (London: TUC Congress House, 2 December 2002).

Harris, A. and Wilson, P. 'Transport sector', paper presented at the Health and Safety Executive *Tackling Work-related Violence – Putting Policies into Practice* conference (London: TUC Congress House, 2 December 2002).

Hatcher, C. and McCarthy, P. 'Workplace bullying: In pursuit of truth in the bully-victim-professional practice triangle', *Australian Journal of Communication,* 29 (3) (2002) 45–58.

Hawken, P. Lovins, A. and Lovins, H. *Natural Capitalism: The Next Industrial Revolution* (London: Earthscan, 1999).

Health and Safety Authority, *Code of Practice – Prevention of Workplace Bullying* (Dublin: Health and Safety Authority, March 2002).

Heilbrun, K. Brock, W. Waite, D. Lanier, A. Schmid, M. Witte, G. Keeney, M. Westendorf, M. Buinavert, L. and Shumate, M. 'Risk factors for juvenile criminal recidivism: The postrelease community adjustment of juvenile offenders', *Criminal Justice and Behaviour,* 27 (3) (2000) 275–91.

Heskett, S. *Workplace Violence: Before, During and After* (Boston: Butterworth-Heinemann, 1996).

Hoel, H. and Cooper, C. *Destructive Conflict and Bullying at Work* (UK: UMIST, Unpublished Report, 2000). See: http://www.le.ac.uk/unions/aut/ umist1.pdf

Hoel, H. and Salin, D. 'Organisational antecedents to workplace bullying', in Einarsen, S. Hoel, H. Zapf, D. and Cooper, C. (eds) *Bullying and Emotional Abuse in the Workplace: International Perspectives in Research and Practice* (UK: Taylor Francis, 2003) 203–18.

Hoel, H. Sparks, K. and Cooper, C. *The Costs Of Violence/Stress At Work And The Benefits Of A Violence/Stress-Free Working Environment* (Manchester, University of Manchester Institute of Science and Technology, report commissioned by the International Labour Organisation, Geneva, 2001).

Hogan, J. 'Your every move will be analysed', *New Scientist* (12 July 2003) 4–5.

Hogarth, M. Gilding, P. and Humphries, R. 'A path to sustainability', in CCH and Freeholds *Australian Master OHS and Environment Guide 2003* (Australia: CCH, 2002) 373–92.

Holston, J. and Appardurai, A. 'Cities and citizenship', *Public Culture*, 8(2) (1996) 187–204.

Homel, R. 'Preventing alcohol-related injuries', in O'Malley, P. and Sutton, A. (eds) *Crime Prevention in Australia: Issues in Policy and Research* (Sydney: The Federation press, 1997) 217–37.

Hornstein, H. *Brutal Bosses and Their Prey: How to Identify and Overcome Abuse in the Workplace* (New York: Riverhead Books, 1996).

House of Representative Standing Committee on Transport, Communications and Infrastructure (HRSCTCI), *Ships of Shame: Inquiry Into Ship Safety* (Canberra: Australian Government Publishing Service, 1992).

Howell, J. (ed.) *Guide for Implementing the Comprehensive Strategy for Serious, Violent, and Chronic Juvenile Offenders* (US Department of Justice, Office of Juvenile Justice and Delinquency prevention, 1995). See: www.ncjrs.org/ pdffiles/guide.pdf

Hume, J. 'Violence grows but cab security ignored', *Security Australia*, 15(11) (1995) 16–8.

Huntington, S. *The Clash of Civilizations and the Remaking of World Order* (New York: Simon and Schuster, 1996).

Income Data Services Ltd (1996) 'Bullying and Harassment at Work', *IDS Employment Law Supplement 76*, (London: IDS, May 1996).

Indermaur, D. *Violent Property Crime* (Sydney: The Federation Press, 1995).

Institute for Social and Ethical Accountability (ISEA) *AA1000 Standard* (London: ISEA, 2000).

International Committee of the Red Cross, *Guantanamo Bay: Overview of the ICRC's Work for Internees*(2003). See: www.icrc.org/web/eng/siteeng0.nsf/html/ 5QRC5V?OpenDocument

International Labour Organisation (ILO), 'Violence at work', *SafeWork* (Geneva: International Labour Office, Report, 1996–2002). See: http://www.ilo.org/ public/english/protection/safework/violence/index.htm

International Labour Organisation (ILO), *Code of Practice on Workplace Violence in Services Sectors and Measures to Combat this Phenomenon* (Geneva: International Labour Organisation, October 2003a).

International Labour Organisation (ILO), *Violence and Stress at Work in Services Sectors*, an ILO Code of Practice, Second draft of in-progress document (Geneva: International Labour Organisation, February 2003b).

International Labour Organisation, International Council of Nurses, World Health Organisation, and Public Services International (ILO/ICN/WHO/PSI), *Joint Programme on Workplace Violence in the Health Sector* (Geneva: ILO, informal technical consultation documents, April 2002a).

International Labour Organisation, International Council of Nurses, World Health Organisation, and Public Services International (ILO/ICN/WHO/PSI), *Framework Guidelines for Addressing Workplace Violence in the Health Sector* (Geneva: ILO, 2002b).

International Crime Victim Survey (ICVS)

International Standards Organisation (ISO). See: www.iso.ch

Ironside, M. and R. Seifert, 'Tackling bullying in the workplace: The collective dimension', in Einarsen, S. and Hoel, H. Zapf, D. and Cooper, C. (eds) *Bullying and Emotional Abuse in the Workplace: International Perspectives in Research and Practice* (UK: Taylor Francis, 2003) 383–99.

Jack, I. 'India's lost talisman', *The Independent* (London, October 26th 1990).

Janoff-Bulman, R. *Shattered Assumptions – Towards a New Psychology of Trauma* (New York: The Free Press, 1992).

Janoff-Bulman, R. 'Assumptive worlds and the stress of traumatic events: Application of the schema construct', *Social Cognition*, 7 (1989) 113–36.

Janoff-Bulman, R. and Freize, I. 'A theoretical perspective for understanding reactions to victimisation', *Journal of Social Issues*, 39(2) (1983) 1–17.

Janoff-Bulmann, R. and Schwartzenberg, S. 'Towards a general model of personal change: Applications to victimisation and psychotherapy', in Snyder, C. and Forsyth, D. (eds) *Handbook of Social and Clinical Psychology: The Health Perspective* (New York: Pergamon, 1990) 488–508.

Jeffery, R. and Zahm, D. 'Crime prevention through environmental design, opportunity theory, and rational choice models', in Clarke, R. and Felson, M. (eds) *Routine Activity and Rational Choice: Advances in Criminological Theory*, vol. 5, (New Jersey: Transaction publishers, 1993).

Jenkins, E. 'Workplace homicide: Industries and occupations at high risk', *Occupational Medicine*, 11(2) (1996) 219–25.

Johnstone, R. 'Paradigm crossed? The statutory occupational health and safety obligations of the business undertaking', *Australian Journal of Labour Law*, 12 (1999) 73–112.

Johnstone, R. *Occupational Health and Safety Law and Policy* (Sydney: The Law Book Company, 1997).

Jones, K. 'The inside story', *CCH OHS Magazine August/September 2003* (Australia, North Ryde, CCH, 2003).

Jurgensmeyer, M. 'Religious terror and global war', in Calhoun, C. Price, P. and Trimmer, A. (eds) *Understanding September 11* (New York: The New Press, 2002).

Kant, I. *Religion Within the Limits of Reason Alone* (New York: Harper and Row, trans. Green, T. and Hudson, H. 1960).

Karpin, D. *Enterprising Nation: Renewing Australia's Managers to Meet the Challenge of the Asia Pacific Century* (Canberra: AGPS, 1995).

Kaufer, S and Mattman, J. *The Cost of Workplace Violence to American Business* (Palm Springs, CA: Workplace Violence Research Institute, 1998).

Kawachi, I. Kennedy, B. and Wilkinson, R. 'Crime: social disorganization and relative deprivation', *Social Science and Medicine*, 48 (1999) 719–31.

Kearns, D. McCarthy, P. and Sheehan, M. 'Organisational restructuring: Considerations for workplace rehabilitation professionals', *Australian Journal of Rehabilitation Counselling*, 3 (1) (1997) 21–9.

Keashly, L. 'Interpersonal and systematic aspects of emotional abuse at work: the target's perspective', *Violence and Victims*, 16 (3) (2001) 233–68.

Keashley, L. and Jagatic, K. 'By any other name: American perspectives on workplace bullying' in S. Einarsen, S. Hoel, H. Zapf, D. and Cooper, C. (eds) *Bullying and Emotional Abuse in the Workplace: International Perspectives in Research and Practice* (London: Taylor & Francis, 2003) 31–62.

Keatsdale Pty Ltd, *A Report on Taxi Driver Safety* (Sydney: New South Wales Department of Transport, 1995).

Kells, P. Keynote address at the *Global Perspectives on Effective Workplace Safety Strategies Symposium* (Melbourne: convention center, 15–16 March 2001). See: http://www.safecommunities.ca

Kennedy, M. 'Systems bullying (the police)', in McCarthy, P. Rylance, J. Bennett, R. and Zimmerman, H. (eds) *Bullying: From Backyard To Boardroom*, 2nd Edition (Sydney: Federation Press, 2001).

Kenny, D. 'Young people in custody health survey: Mental health, cognitive and academic findings', paper presented to the *Juvenile Justice: From the Lessons of the Past to a Road Map of the Future Conference* (Sydney: The Australian Institute of Criminology in conjunction with the NSW Department of Justice, Citigate Sebel Hotel, 1–2 December 2003).

Kivimäki, K. Elovainio, M. and Vathera, J. 'Workplace bullying and sickness absence in hospital staff', *Occupational and Environmental Medicine*, 57 (2000) 656–60.

Kivimaki, M. Elovainio, M. Vahtera, J. and Ferrie, J. 'Organisational justice and health of employees: prospective cohort study', *Occupational and Environmental Medicine*, 60 (2003) 27–34.

Lacan, J. *The Four Fundamental Concepts of Psychoanalysis* (New York: W.W. Norton & Co, edited by Milleer, J-A and trans A. Sheridan, 1977).

Lane, T. *The Pressing Need to Explore the Neglected World of the Seafarer* (London: Nautical Year Book, Lloyds, 1999).

LeBlanc, M. and Kelloway, K. 'Predictors and outcomes of workplace violence and aggression', *Journal of Applied Psychology*, 87 (2002) 444–53.

Lennane, J. 'Bullying in medico-legal examinations', in McCarthy, P. Sheehan, M. and Wilkie, W. (eds) *Bullying: From Backyard To Boardroom* (Sydney: Millennium, 1996) 97–118.

Lerner, M. 'The desire for justice and reactions to victims', in Macaulay, J. and Berkowitz, L. (eds) *Altruism and Behaviour: Social Psychological Studies of Some Antecedents and Consequences* (New York: Academic Press, 1970) 205–9.

Lester, A. 'Crime reduction through product design', *Trends and Issues in Crime and Criminal Justice*, number 206 (Canberra: Australian Institute of Criminology, 2001).

Lewis, D. 'Workplace bullying – A case of moral panic?', in Sheehan, M. Ramsay, S. and Patrick, J. (eds) *Transcending Boundaries: Integrating People, Processes and Systems* (Brisbane: Griffith University School of Management, Conference Proceedings 6–8 September 2000).

Leymann, H. *Explanation of the Operation of the LIPT – Questionnaire (Leymann Inventory of Psychological Terrorisation)* (Brisbane: Griffith University, School of Management, translated by Zimmermann, H. 1997).

Leymann, H. 'The content and development of mobbing at work', *European Journal of Work and Organizational Psychology*, 5(2) (1996) 165–84.

Leymann, H. *Vuxenmobbning pa[o] Svenska Arbetsplatser. En Rikstäckande Undersökning med 2.428 Intervjuer* (Mobbing at Swedish workplaces – a study of 2428 individuals) (Stockholm: Arbetarskyddsstyrelsen, Delrapport, 1992a).

Leymann, H. *Från Mobbing till Utslangning I Arbetslivet* (From mobbing to expulsion in work life) (Stockholm: Publica, 1992b).

Leymann, H. *Manlight vid Vuxen Mobbing. En Rikstäckande Undersökning med 2.428 Intervjuer* (Gender and mobbing – a study of 2428 individuals) (Stockholm: Arbetarskyddsstyrelsen, Delrapport 2, 1992c).

Leymann, H. *Psykiatriska Hälsppproblem I Samband med Vuxenmobbning. En Rikstäckande Undersökning med 2.428 Intervjuer* (Psychiatric problems after mobbing – a study of 2428 individuals) (Stockholm: Arbetarskyddsstyrelsen, Delrapport 3, 1992d).

Leymann, H. *Handbok för Anvandning av LIPT – Formuläret for Kartläggning av Risker för Psykiskt va[o]ld I Arbetsmiljön* (The LIPT questionnaire – a manual). (Stockholm: Violen, 1990a).

Leymann, H. 'Mobbing and psychological terror at workplaces', *Violence and Victims*, 5 (1990b) 119–25.

Leymann, H. *Ingen Annan Utväg* (no way out) (Stockholm: Wahlstrom and Widstrand, 1988).

Leymann, H. *Vuxenmobbning – Om Psykiskt Våld I Arbetslivet* (Mobbing – psychological violence at work places) (Lund: Studentlitteratur, 1986).

Leymann, H. and Gustafsson, A. 'Mobbing and the development of post-traumatic stress disorders', *European Journal of Work and Organizational Psychology*, 5(2) (1996) 251–76.

Liefooghe, A. and Davey, K. 'Accounts of workplace bullying: the role of the organization', *European Journal of Work and Organizational Psychology*, 10 (4) (2001) 375–92.

Liefooghe, A., and Olafsson, R. 'Scientists and amateurs: Mapping the bullying domain', *International Journal of Manpower*, 20(1/2) (1999) 39–49.

Limerick, D. and Cunnington, B. *Managing the New Organisation: A Blueprint for Networks and Strategic Alliances* (Chatswood: Business and Professional publishing, 1993).

Littler, C. 'Downsizing: A disease or a cure', *HR Monthly* (August 1996) 8–12.

Long Island Coalition for Workplace Violence Awareness and Prevention, *Workplace Violence Awareness and Prevention: an Information and Instructional Package for use by Employers and Employees* (United States: Long Island Coalition, 1996).

Lynch, M. Buckman, J. and Krenske, L. 'Youth justice: criminal trajectories', *Trends and Issues in Crime and Criminal Justice*, number 265 (Canberra: Australian Institute of Criminology, 2003).

Lyotard, J-F. *The Differend: Phrases in Dispute* (Minneapolis: University of Minnesota Press, 1988).

Lyotard, J-F. *The Postmodern Condition: A Report on Knowledge* (Manchester: Manchester University Press, 1984).

Maffesoli, M. *Le Temps Des Tribus* (Paris: KlincKseick, 1988).

Maggio, M. 'Keeping the workplace safe: A challenge for managers', *Federal Probation*, 60(1) (1996).

Mailer, N. 'Only in America', *New York Review of Books*, 27th March 2003. See: http://www.nybooks.com/articles/16166

Makkai, T. and McGregor, K. *Drug Use Monitoring in Australia: 2002 Annual Report on Drug Use Among Police Detainees*, Research and Public Policy Series number 47 (Canberra: Australian Institute of Criminology, 2003).

Malone, K. and Hasluck, L. 'Geographies and exclusion: young people's perceptions and their use of public space', *Family Matters*, 49 (1998) 20–26.

Manne, D. 'Excision crosses ethical border', See: www.thecouriermail.com.au (6th November 2003)19.

Mapstone, E. *War of Words: Men and Women Arguing* (London: Chatto and Windus, 1998).

Marosi, R. 'One of the most dangerous jobs in New York: Gypsy cab driver', *Columbia University News Service*, (1996). See: www.taxi-1.org/marosi.htm

Marsh, J. and Moohan, J. 'Small firms in rural areas', *Security Journal*, 16(3), (2003) 39–49.

Martin, C. 'Crime and control in Australian urban space', *Current Issues in Criminal Justice*, 12(1) (2000), 79–92.

Maskell-Knight, H. 'Researching the literature: the epidemiology of violence in rural and remote Australia', paper presented at the *Evaluation in Crime and Justice: Trends and Methods conference* (Canberra: Australian Institute of Criminology, 24–25 March 2003).

Matthiesen, S. and Einarsen, S. 'MMPI-2-configurations among victims of bullying at work', *European Journal of Work and Organizational Psychology*, 10 (2001) 467–84.

Mayhew, C. 'Occupational violence and prevention strategies', *Master OHS and Environment Guide* (North Ryde: CCH Australia, 2003a) 547–69.

Mayhew, C. 'Occupational violence: a neglected OHS issue?' *Policy and Practice in Health and Safety*, 1(1) (2003b) 31–58.

Mayhew, C. 'Preventing violence against health workers', paper presented at WorkSafe Victoria seminar on occupational violence (Melbourne, 13 June 2003c) (published on line see: www.workcover.vic.gov.au).

Mayhew, C. 'Occupational violence in industrialized countries: types, incidence patterns, and 'at risk' groups of workers', in M. Gill, B. Fisher and V. Bowie (eds) *Violence at Work: Causes, Patterns and Prevention* (United Kingdom: Willan Publishing, 2002a) 21–40.

Mayhew, C. 'Getting the message across to small business about occupational violence and hold-up prevention: a pilot study', *Journal of Occupational Health and Safety – Australia and New Zealand*, 18(3) (2002b) 223–30.

Mayhew, C. 'Suicide and seafarers: a pilot study', *Journal of Occupational Health and Safety – Australia and New Zealand*, 18(4) (2002c) 375–81.

Mayhew, C. 'The detection and prevention of cargo theft', *Trends and Issues in Crime and Criminal Justice*, no. 214 (Canberra: Australian Institute of Criminology, 2001a).

Mayhew, C. 'Pilot evaluation of the small business self-audit checklist in Tasmania' (Canberra: Australian Institute of Criminology, unpublished report, 2001b).

Mayhew, C. 'Violent assaults on taxi drivers: incidence patterns and risk factors', *Trends and Issues in Crime and Criminal Justice*, no. 178 (Canberra: Australian Institute of Criminology, 2000a).

Mayhew, C. *Preventing Client-Initiated Violence: A Practical Handbook*, Research and Public Policy series no. 30 (Canberra: Australian Institute of Criminology, 2000b).

Mayhew, C. *Preventing Violence Within Organisations: A Practical Handbook*, Research and Public Policy series, no. 29 (Canberra: Australian Institute of Criminology, 2000c).

Mayhew, C. *Violence in the Workplace-Preventing Commercial Armed Robbery: A Practical Handbook*, Research and Public Policy series No. 33 (Canberra: Australian Institute of Criminology, 2000d).

Mayhew, C. 'Preventing assaults on taxi drivers in Australia', *Trends and Issues in Crime and Criminal Justice*, no. 179, (Canberra: Australian Institute of Criminology, 2000e).

Mayhew, C. 'Occupational violence: A case study of the taxi industry', in Mayhew, C. and Peterson, C. *Occupational Health and Safety in Australia:*

Industry, Public Sector and Small Business (Sydney: Allen and Unwin, 1999a) 127–39.

Mayhew, C. 'Identifying patterns of injury in small business: piecing together the data jigsaw', in Mayhew, C. and Peterson, C. *Occupational Health and Safety in Australia: Industry, Public Sector and Small Business,* (Sydney: Allen and Unwin, 1999b) 105–15.

Mayhew, C. 'Occupational health and safety management systems (OHSMS) impact on OHS performance in the franchised outlets of a large fast-food chain', paper presented at the *Voluntary Guidelines for Management Systems for the Working Environment* workshop (Brussels: unpublished proceedings, 14–16 June 1999c).

Mayhew, C. *Barriers to Implementation of Known OHS Solutions in Small Business* (Canberra: National Occupational Health and Safety Commission and Division of Workplace Health and Safety, AGPS, 1997).

Mayhew, C. and Chappell, D. *The Occupational Violence Experiences of Some Australian Health Workers: An Exploratory Study,* published as monograph in special issue of *The Journal of Occupational Health and Safety – Australia and New Zealand,* 19(6) (North Ryde: CCH Australia, December 2003).

Mayhew, C. and Grewal, D. 'Occupational Violence/Bullying in the maritime industry: A pilot study', *The Journal of Occupational Health and Safety – Australia and New Zealand,* 19 (5) (2003) 457–63.

Mayhew, C. and McCarthy, P. *A Pilot Study of the Occupational Violence Experiences of Juvenile Justice Workers* (Brisbane: Griffith University, School of Management, unpublished confidential report, November 2003).

Mayhew, C. McCarthy, P. Barker, M. and Sheehan, M. 'Student aggression in tertiary education institutions', *Journal of Occupational Health and Safety – Australia and New Zealand,* 18(4) (2003) 327–35.

Mayhew, C. McCarthy, P. Chappell, D. Quinlan, M. Barker, M. and Sheehan, M. 'Measuring the extent of impact from occupational violence and bullying on traumatised workers', *Employee Rights and Responsibilities Journal* special issue on 'The traumatized worker' (in press, 2004).

Mayhew, C. and Quinlan, M. 'Occupational violence in long distance road transport: A study of 300 Australian truck drivers', *Current Issues in Criminal Justice,* 13 (1) (2001) 36–46.

Mayhew, C. and Quinlan, M. *Occupational Health and Safety Amongst 300 Long Distance Truck Drivers: Results of an Interview-Based Survey,* research report conducted as part of the Motor Accident Authority of New South Wales *Safety Inquiry into the Long Haul Trucking Industry* (Sydney: University of New South Wales, Industrial Relations Research Centre, 2000).

Mayhew, C. and Quinlan, M. 'The relationship between precarious employment and patterns of occupational violence: Survey evidence from thirteen occupations', in Isaksson, K. Hogstedt, C. Eriksson, C. and Theorell, T. (eds) *Health Effects of the New Labour Market,* (New York: Kluwer Academic/Plenum publishers, 1999) 183–205.

Mayhew, C. Quinlan, M. and Bennett, L. *The Effects of Subcontracting/ Outsourcing on Occupational Health and Safety* (Sydney: University of New South Wales, Industrial Relations Research Centre, 1996).

Mayhew, P. 'Counting the costs of crime in Australia', *Trends and Issues in Crime and Criminal Justice,* no. 247 (Canberra: Australian Institute of Criminology, April 2003)

McCarthy, P. 'Bullying: A postmodern experience', in Einarsen, S. Hoel, H. Zapf, D. and Cooper, C. (eds) *Bullying and Emotional Abuse in the Workplace: International Perspectives in Research and Practice* (London: Taylor and Francis, 2003a).

McCarthy, P. 'A letter to HR managers on workplace bullying', The *Journal of Occupational Health and Safety – Australia and New Zealand,* 19(4) (2003b) 307–11.

McCarthy, P. 'New research directions: Implications for implementing anti-bullying policies', in *Workplace Bullying Symposium, Constructive Leadership: Making Anti-Bullying Initiatives Work Proceedings* (Brisbane: Griffith University, 21 March 2002).

McCarthy, P. 'The bullying syndrome: Complicity and responsibility', in McCarthy, P. Rylance, J. Bennett, L. and Zimmermann, H (eds) *Bullying: From Backyard To Boardroom,* second edition (Sydney: Federation Press, 2001).

McCarthy, P. 'The bully-victim at work', in Sheehan, M. Ramsay, S. and Patrick, J. (eds) *Transcending Boundaries: Integrating People, Processes and Systems* (Brisbane: Griffith University, proceedings of the 2000 conference, September 2000a) 251–6.

McCarthy, P. 'Prefazione to Antonio Casilli', *Stop Mobbing* (Roma: Derive Approdi, 2000b).

McCarthy, P. 'Strategies between managementality and victim-mentality in the pressures of continuous change', in Fraser, C. Barker, M. and Martin, A. (eds) *Organisations Looking Ahead: Challenges and Directions* (Brisbane: Griffith University, School of Management, conference proceedings November 22–23, 1999) 211–19.

McCarthy, P. 'Guidelines: Maintaining vital people and viable workplaces – beyond victimisation – in the pressures of continuous change', *Conflict and Violence in the Workplace* (Canberra: Consolidated Conferences and the Australian Psychological Society LTD, Rydges Lakeside, April 23–24, 1998a) 8–19.

McCarthy, P. and Barker, M. 'Workplace bullying risk audit' *Journal of Occupational Health and Safety, Australia and New Zealand,* vol. 16(5) pp. 409–418.

McCarthy, P. 'Consulting-violence in city revitalisation', McCarthy, P. Sheehan, M. Wilkie, S. and Wilkie, W. *Bullying: Causes, Costs and Cures* (Brisbane: Beyond Bullying Association, Inc, 1998b).

McCarthy, P. *Postmodern Desire: Learning From India* (New Delhi: Promilla publishers, 1994).

McCarthy, P. 'Postmodern pleasure and perversity: Scientism and sadism', in Amiran, E. and Unsworth, J. (eds) *Essays in Postmodern Culture* (New York: Oxford University Press, 1993).

McCarthy, P. Barker, M. Sheehan, M. and Henderson, M. 'Workplace bullying' in *Equal Opportunity Training Manual* (Sydney: CCH, 2001).

McCarthy, P. Henderson, M. Barker, M. and Sheehan, M. *Socially Responsible Management and Ethical Investment: Employment Practices* (Brisbane: Griffith University, Centre for Research on Employment and Work (CREW), 2002a).

McCarthy, P. Henderson, Sheehan, M. and Barker, M. 'Workplace Bullying' in CCH and Freeholds *Australian Master OHS and Environment Guide 2003* (Australia: CCH, 2002b) 519–46.

McCarthy, P. Henderson, M. Sheehan, M. and Barker, M. 'Socially responsible management and ethical investment: Recent trends and industry implications' (Brisbane: Griffith University, Centre for Research on Employment and Work, in press 2004b)

McCarthy, P. and Mayhew, C. 'Workplace bullying/violence in public service organisations', in *Workplace Bullying* conference proceedings (Brisbane: Queensland Working Women's Service, 16–17 October 2003).

McCarthy, P. Mayhew, C. Barker, M. and Sheehan, M. 'Occupational violence in tertiary education: risk factors, perpetrators and prevention', The *Journal of Occupational Health and Safety – Australia and New Zealand*, 19(4) (2003b) 319–26.

McCarthy, P. and Rylance, J. 'The bullying challenge', in McCarthy, P. Rylance, J. Bennett, R. and Zimmerman, H. (eds) *Bullying: From Backyard To Boardroom*, 2nd edition (Sydney: Federation Press, 2001).

McCarthy, P. Sheehan, M. Barker, M. and Henderson, M. 'Ethical investment and workplace bullying: Consonances and dissonances', in Lewis, D. and Sheehan, M. (eds) *International Journal of Management and Decision Making Special Edition on Workplace Bullying*, 4(1) (2003a) 11–23.

McCarthy, P. Sheehan, M. and Kearns, D. *Managerial Styles and Their Effects on Employees' Health and Well-Being in Organisations Undergoing Restructuring* (Brisbane: Griffith University, School of Organisational Behaviour and Human Resource Management, 1995).

McCarthy, P. Sheehan, M. and Wilkie, W. (eds) *Bullying: From Backyard to Boardroom*, second edition (Sydney: Federation Press, 2001).

McCarthy, P. Sheehan, M. and Wilkie, W. (eds) *Bullying: From Backyard to Boardroom*, (Sydney: Millenium Books, 1996).

McCarthy, P. and Zimmermann, H. 'Self-help needs of persons experiencing bullying at work' (Brisbane: Griffith University, School of Management, unpublished survey results, 2002).

McCarthy, P. Barker, M. Brown, K. Powell, D. 'Cross -sector partnerships: Limitations of third-way solution to disadvantage', unpublished article submitted for review 2003.

McCulloch, J. 'Counter-terrorism, human security and globalisation – from welfare to the warfare state', *Current Issues in Criminal Justice*, 14(3) (2003) 283–98.

McPherson, A. and Hall, W. 'Psychiatric impairment, physical health and work values among unemployed and apprenticed young men', *Australian and New Zealand Journal of Psychiatry*, 17 (1983) 335–40.

Megalogenis, G. 'The million voters who matter', See: www.Inquirer@theaustralian.com.au (2002), p.17 & 20.

Metcalf, B. 'Traditionalist Islamic activism: Deoband, tablighis, and talibs', in Calhoun, C. Price, P. and Trimmer, A. (eds) *Understanding September 11* (New York: The New Press, 2002).

Mikkelsen, E. *Mobning i Arbejdslivet: Hvorfor og for Hvem er den så Belastende?* (Workplace bullying: Why and for whom is bullying such a strain?) *Nordisk Psykologi*, 53(2) (2001a) 109–31.

Mikkelsen, E. 'Workplace bullying: It's prevalence, aetiology and health correlates' (Aarhus: University of Aarhus, Department of Psychology, PhD thesis, 2001b).

Mikkelsen, E. and Einarsen, S. 'Basic assumptions and symptoms of post-traumatic stress among victims of bullying at work', *European Journal of Work and Organisational Psychology*, 11(1) (2002) 87–111.

Milgram, S. *Human Relations*, 18 (1965) 56–76.

Mirrlees-Black, C. and Ross, A. *Crime Against Retail and Manufacturing Premises: Findings from the 1994 Commercial Victimisation Survey* (London: Home Office, 1995).

Moore, M. (2001) *Stupid White Men – and Other Sorry Excuses for the State of the Nation*, Regan Books, New York.

Morgan, F. *Road Rage: Driving Related Violence in Western Australia* (Perth: The University of Western Australia, Crime Research Centre, report for the Royal Automobile Club of Western Australia, 1997).

Morgan, R. *The Demon Lover: The Roots of Terrorism* (New York: Washington Square press, 2001).

Morris, A. Reilly, J. Berry, S. and Ransom, R. *The New Zealand National Survey of Crime Victims 2001* (Wellington, New Zealand: Ministry of Justice, May 2003).

Mullen, E. 'Workplace violence: cause for concern or the construction of a new category of fear', *Journal of Industrial Relations*, 39(1) (1997) 21–32.

Mumby, D. and Stohl, C. 'Disciplining organizational communication studies', *Management Communication Quarterly*, 10(1) (1996) 5–72.

Muscat, G. and Carcach, C. *Australian Crime: Facts and Figures* (Canberra: Australian Institute of Criminology, 2002).

Myers, D. 'A workplace violence prevention planning model', *Journal of Security Administration*, 19(2) (1996) 1–19.

National Board of Occupational Safety and Health, *Victimization at Work* (Stockholm: National Board of OHS, 1993).

National Health Service, *A Safer Place to Work: Protecting NHS Hospital and Ambulance Staff from Violence and Aggression* (London: The Stationary Office, report by the Comptroller and Auditor General, 27 March 2003). See: www.hse.gov.uk/healthservices/violence.htm

National Health Service (NHS), *Withholding Treatment from Violent and Abusive Patients in NHS Trusts: We Don't Have To Take This Resource Guide* (UK: National Health Service, Department of Health, 2002). See: www.nhs.gov.uk

National Health Service (NHS), *Managing Violence in Ambulance Trusts: We Don't Have To Take This Resource Guide* (UK: National Health Service, Department of Health, 2000).

National Health Service (NHS), *Effective Management of Security in A&E* (UK: NHS Executive, 1997).

National Health and Medical Research Council (NH&MRC), *When It's Right in Front of You: Assisting Health Care Workers to Manage the Effects of Violence in Rural and Remote Australia* (Canberra: NH&MRC, 2002). See: www.nhmrc.gov.au/publications.

National Institute for Occupational Safety and Health (NIOSH), 'Proposed data collections submitted for public comment and recommendations: Developing communication to reduce workplace violence and assault against taxicab drivers' (Atlanta: Centers for Disease Control and Prevention, 2000). See: www.cdc.gov/niosh/00–3061.html

National Institute for Occupational Safety and Health (NIOSH) 'Preventing homicide in the workplace: Workers in certain industries and occupations are at increased risk of homicide', *NIOSH Alert* publication no. 93–109 (Atlanta: National Institute for Occupational Safety and Health, CDC, 1995).

National Occupational Health and Safety Commission (NOHSC), *Compendium of Workers' Compensation Statistics, Australia 2000–2001* (Canberra: NOHSC, December 2002). See: www.nohsc.gov.au/statistics/publications.

National Occupational Health and Safety Commission (NOHSC), *Work-Related Traumatic Fatalities in Australia, 1989 to 1992*, written by Driscoll, T. Mitchell, R. Mandryk, J. Healey, S. and Hendrie, L. (Canberra: Ausinfo, 1999).

Neuman, W. *Social Research Methods: Qualitative and Quantitative Approaches* (Massachusetts: Allyn and Bacon, 1994).

Neuman, J. and Baron, R. 'Social antecedents of bullying: a social interactionist perspective', in Einarsen, S. Hoel, H. Zapf, D. and Cooper, C. *Bullying and Emotional Abuse in the Workplace* (London: Taylor and Francis, 2003) 185–202.

Nicarthy, G. Gottlieb, N, and Goffman, S. *You Don't Have to Take It! A Women's Guide for Confronting Emotional Abuse at Work* (Washington DC: Seal Press, 1993).

Niedl, K. 'Mobbing and well-being: Economic and personnel implications', *European Journal of Work and Organisational Psychology*, 5 (1996) 239–49.

Niedl, K *Mobbing/bullying am Arbeitsplatz. Eine Empirische Analyse zum Phanomen Sowie zu Personalwirtschaftlich Relevanten Effekten von Systematis-chen Feindseligkeiten* (Mobbing/bullying at work. An empirical analysis of the phenomenon and of the effects of systematic harassment on human resource management) (Munich: Hampp, 1995).

Norris, F. Byrne, C. Diaz, E. Kaniastry, K. *The Range, Magnitude, and Duration of Effects of Natural and Human-Caused Disasters: A Review of the Empirical Literature* (Boston, Mass: National Centre for PTSD, 2002).

North, C. Nixon, S. Shariat, S. Mallonee, S. McMillen, C. Spitznagel, E. Smith, E. 'Psychiatric disorders amongst survivors of the Oklahoma City bombing', *JAMA*, 282 (1999) 755–62.

Northern Territory Parliament of Australia, *Public Order and Anti-Social Conduct Act* (serial number 335) (Darwin, Government of the Northern Territory, 2001).

Occupational Health and Safety Administration (OSHA), *Recommendations for Workplace Violence Prevention Programs in Late-Night Retail Establishments* (Washington DC: United States Department of Labor, 1998).

Office of the Employee Ombudsman, *Bullies Not Wanted: Recognising and Eliminating Bullying in the Workplace* (Adelaide: Office of the Employee Ombudsman of South Australia, 2000)

Office of the Public Sector Merit and Equity, Queensland (OPSME), *Grievance Directive* (Brisbane: OPSME, July 2003).

Oliver, C. 'Employers responsible for suicide', *CCH OHS Magazine* (October/ November 2001) 32.

Ore, T. 'Differences in assault rates among direct care workers', *The Journal; of Occupational Health and Safety – Australia and New Zealand*, 19(3) (2003) 225–33.

Painter, K. and Farrington, D. 'The crime reducing effect of improved street lighting: The Dudley project', in Clarke, R. (ed.) *Situational Crime Prevention Successful Case Studies* second edition (New York: Harrow and Heston, 1997).

Paoli, P. 'Third European survey on working conditions', cited in *SafeWork: Violence at Work in the European Recent Finds* (Geneva: International Labour Organization, 2000). See: www.ilo.org/public/english/protection/safework/ violence/eusurvey/eusurvey.htm

Parker, A. Hubinger, L. Green, S. Sargent, L. and Boyd, R. *A Survey of the Health, Stress and Fatigue of Australian Seafarers* (Canberra: report written by staff of the Queensland University of Technology for the Australian Maritime Safety Authority, 1997).

Paterson, B. and Leadbetter, D. 'Standards for violence management training', in Gill, M. Fisher, B. and Bowie, V. (eds) *Occupational Violence in Industrialised Countries* (Cullumpton: Willan Press, 2002) 132–50

Perrone, S. 'Crimes against small business in Australia: a preliminary analysis', *Trends and Issues in Crime and Criminal Justice,* number 184 (Canberra: Australian Institute of Criminology, 2000).

Perrone, S. *Violence in the Workplace,* Research and Public Policy series report no. 22 (Canberra: Australian Institute of Criminology, 1999).

Peters, T. *Thriving on Chaos: Handbook for a Managerial Revolution* (New York: Alfred A. Knopf, 1987).

Peterson, C. 'Stress among health care workers', *Journal of Occupational Health and Safety – Australia and New Zealand,* 18 (1) (2003) 49–58.

Polk, K. 'Youth violence: myth and reality', in Chappell, D. and Egger, S. (eds) *Australian Violence Contemporary Perspectives II* (Canberra: Australian Institute of Criminology, 1995).

Portman, J. *When Bad Things Happen To Other People* (London: Routledge, 2000).

Poynting, S. 'Bin Laden in the suburbs: attacks on Arab and Muslim Australians before and after 11 September, 2002', *Current Issues in Criminal Justice,* 14(1) (2002) 43–64.

Pusey, M. 'What's wrong with economic rationalism?' in Horne, D. (ed.) *The Trouble with Economic Rationalism* (Clayton: Scribe, 1992) 63–9.

Queensland Parliament, *Workplace Health and Safety Act 1995* (Brisbane: Government printer, Act no. 25 of 1995).

Queensland Workplace Bullying Taskforce, *Creating Fair and Safe Workplaces: Strategies to Address Workplace Harassment* (Brisbane: Department of Industrial Relations, report of Taskforce, 2002).

Quinlan, M. Mayhew, C. and Bohle, P. 'The global expansion of precarious employment, work disorganisation and consequences for occupational health: a review of recent research', *International Journal of Health Services,* 31(2) (2001) 335–414.

Quinlan, M. and Mayhew, C. 'Precarious employment and workers' compensation', *International Journal of Law and Psychiatry,* 22(5–6) (1999) 491–520.

Rayner, C. 'Building a business case for tackling bullying in the workplace: beyond a cost-benefit approach', in Sheehan, M. Ramsay, C. and Patrick, J. (eds) *Transcending Boundaries,* (Brisbane: conference proceedings, September, 2000) 26–31.

Reinhold, B. *Toxic Work: How to Overcome Stress, Overload, and Burnout and Revitalise Your Career* (New York: Plume, 1997).

Rhaman, M. and Rattanani, L. 'Savagery in Bombay', *India Today* (New Delhi: Living India, January 31, 1998).

Roberts, S. *Occupational Mortality Among Merchant Seafarers in the British, Singapore and Hong Kong Fleets (1981–1995)* (Cardiff: University of Wales, Seafarers International Research Centre, 1998).

Robinson, H. *Rage* (Port Melbourne, Australia: Lothian, 1999).

Romei, S. 'My bosses were swindlers says Enron whistleblower', *The Weekend Australian* (31st July 2002) 13. See: http://www.theaustralian.com.au

Rose, G. *The Broken Middle* (Oxford: Blackwell, 1992).

Rose, G. 'Architecture to philosophy – The postmodern complicity', *Theory Culture and Society* 5(2–3) (1988) 357–72.

Rylance, J. *Reaffirmation Processes: A Study Of The Experience Of Responding To Workplace Abuse* (Brisbane: University of Queensland, School of Social Work and Social Policy, unpublished PhD Thesis, 2001).

Sade, Marquis de (1968) *Juliette* (New York: Grove, trans by Warinhouse, A. 1968).

Said, E. *Orientalism* (New York: Vintage, 1979).

Salecl, R. 'Spoils of freedom', in Zournazi, M. (ed.) *Foreign Dialogues* (Annandale: Pluto Press, 1998).

Salkin, S. 'Safe and secure?' *Warehousing Management* 10 (December 1999). See: http://www.manufacturing.net/wm/index.asp?layout=articleWebzine&articleid=CA121823

Salzano, J. and Hartman, S. 'Cargo crime', *Transnational Organized Crime*, 3(1) (1997) 39–49.

Sarkissian Associates and ACT Planning and Land Management, *Australian Capital Territory Crime Prevention and Urban Design Resource Manual* (Canberra: ACT Department of Urban Services, 2000).

Suzuki, D. and Dressel, H. *Good News For A Change: Hope for a Troubled Planet* (Australia: Allen & Unwin, 2002).

Schlenger, W. Caddell, J. Ebert, L. Jordan, B. Rourke, K. Wilson, D. Thalji, L. Dennis, J. Fairbank, J. and Kulka, R. 'Psychological reactions to terrorist attacks: Findings from the national study of American's reactions to September 11', *JAMA*, 228(5) (2002) 581–8.

Scholz J. and Gray W. 'OHSA enforcement and workplace injuries: a behavioural approach to risk assessment', *Journal of Risk and Uncertainty*, 3 (1990) 283–305.

Schuster, M. Stein, B. Jaycox, L. Collins, R. Marshall, G. Elliott, M. Zhou, A. Kanovse, D. Morrison, J. and Berry, S. (2001) 'A national survey of stress reactions after the September 11, 2001 terrorist attacks', *New England Journal of Medicine*, 345(20) (2001) 1507–12.

Scott, M. and Stradling, S. 'Post-traumatic stress disorder without the trauma', *British Journal of Clinical Psychology*, 33 (1994) 17–31.

Seabrook, J. *The No-nonsense Guide to Class, Caste and Hierarchies* (London: New Internationalist and Verso, 2002).

Segessenmann, T. *International Comparisons of Recorded Violent Crime Rates for 2000* (Wellington, New Zealand: Ministry of Justice, 2002).

Shannon, S and Fisher, B. 'Workplace violence in the USA: are there gender differences?' in Gill, M. Fisher, B. and Bowie, V. (eds) *Occupational Violence in Industrialised Countries* (Cullumpton: Willan Press, 2002) 90–113.

Sheehan, M. McCarthy, P. Barker, M. and Henderson, M. 'Ethical investment and workplace bullying: consonances and dissonances', in Lewis, D. and Sheehan, M. (eds) *International Journal of Management and Decision Making*, special edition on workplace bullying (2002).

Sheehan, M. McCarthy, P. Barker, M, and Henderson, M. 'A model for assessing the impacts and costs of workplace bullying' (Dublin: Paper presented at the Standing Conference on Organisational Symbolism (SCOS), Trinity College, June 30 to July 4, 2001).

Sherman, L. Farrington, B. Welsh, B. and Layton MacKenzie, D. (eds) *Evidence-based Crime Prevention* (London: Routledge, 2002).

Showalter, E. *Hystories: Hysterical Epidemics and Modern Culture* (New York: Columbia University Press, 1997).

Simmons, A. *Territorial Games: Understanding and Ending Turf Wars at Work* (New York: AMACOM, 1998).

Singer, P. *One World: The Ethics of Globalisation* (Melbourne: The Text Publishing company, 2002).

Small, S. 'Secure measures', *Warehousing Management,* 6 (June 1998). See: http://www.manufacturing.net/wm/index.asp?layout=articleWebzine&articleid=CA149032

Smith, D. *The Atlas of War and Peace* (London: Earthscan, 2003).

Smith, D. Christianson, E. Vincent, R. and Hanne, N. 'Population effects of the bombing of Oklahoma City', *Journal of the Oklahoma State Medical Association,* 92 (1999) 193–8.

Social Accountability International, *Social Accountability 8000 International Standard* (2001). See: www.cepaa

Social Venture Network, *Standards of Corporate Social Responsibility* (San Francisco: SVN, 1999). See: www.svn.org

Somerville, P. 'On guard', *National Safety* (The Journal of the National Safety Council of Australia, June 2001a) 50–2.

Somerville, P. 'On the move', *National Safety* (May 2001b) 24–9.

Standing Committee on Law and Justice, *Crime Prevention Through Social Support* (Sydney: Parliamentary paper no. 437, NSW Parliament, second report, 2000).

Standing, H. and Nicolini, D. *Review of Workplace-Related Violence* (United Kingdom: report for the Health and Safety Executive, No. 143/1997, Tavistock Institute, 1997).

Stenning, P. *Fare Game, Fare Cop: Victimization of, and Policing by, Taxi Drivers in Three Canadian Cities* (Ottawa: Department of Criminal Justice, 1996).

Stevenson, C. 'Homicides and Violence', *Human Services Standard Book* (Canberra: Australian Institute of Health and Welfare, 19 October 2001). See: http://www.aihw.gov.au/bod/bod_yld_by_disease/t_u_injuries/u2_homicide_and_violence

Swanton, B. and Webber, D. *Protecting Counter and Interviewing Staff from Client Aggression* (Canberra: Australian Institute of Criminology, 1990).

Swedish National Board of Occupational Safety and Health, *Swedish Code (Ordinance AFS 1993:17)* (Stockholm: National Board of Occupational Safety and Health, 1993).

Task Force on the Prevention of Workplace Bullying, *Dignity at Work – The Challenge of Workplace Bullying* (Dublin: The Stationary Office, 2001).

Tattum, D. and Herdman, G. *Bullying: A Whole Prison Response* (Cardiff: Cardiff Institute of Higher Education, Countering Bullying Unit, 1995).

Taylor, C. 'Death on the high seas', *The Sunday Mail* (3/9/2000) 16–8.

Taylor, N. and Mayhew, P. 'Robbery against service stations and pharmacies: recent trends', *Trends and Issues in Crime and Criminal Justice,* no. 223 (Canberra: Australian Institute of Criminology, 2002).

Thompson, C. *Workplace Bullying Project* (Adelaide: Report from Working Women's Centre of South Australia, 1997).

Toohey, P. 'Party animals', *The Weekend Australian Magazine* (12–13 July 2003) 16–9.
Trades Union Congress. *Protect Us From Harm: Preventing Violence at Work* (London: TUC Health and Safety Unit, report by Julia Gallagher, 1999).
Turnbull, J. and Paterson, B. (eds) *Aggression and Violence: Approaches to Effective Management* (London: MacMillan, 1999).
UNISON, *UNISON Members Experience of Bullying at Work* (London: UNISON, 1997).
United Nations Surveys on Crime Trends and Operations of the Criminal Justice System (Geneva: United Nations, 2004a). See: http://www.uncjin.org/ Statistics/WCTS/wets.html
United Nations, *UN Action Against Terrorism* (Geneva: United Nations, 2004b). See:www.un.org/terrorism/
United Nations, *Rio Declaration on Environment and Development* (Brazil: The United Nations Conference on Environment and Development, 1992). See: www.ig.apc.org/habitat/agenda21/rio-dec.html
University of Queensland, *Anti-bullying Policy* (Brisbane: University of Queensland, 2002).
Vandenbos, G. and Bulatao, E. (eds) *Violence on the Job: Identifying Risks and Developing Solutions* (Washington DC: American Psychological Association, 1996).
Warshaw, L. and Messite, J. 'Workplace violence: preventive and interventive strategies', *Journal of Occupational and Environmental Medicine,* 38 (10) (1996) 993–1005, p.999.
Watts, P. and Lord Holme *Corporate Social Responsibility: Meeting Changing Expectations* (Geneva: World Business Council of Sustainable Development, 1998). See: http://www.wbcsd.ch
Wei, Z. Makkai, T. and McGregor, K. 'Drug use among a sample of juvenile detainees', *Trends and Issues in Crime and Criminal Justice,* number 258 (Canberra: Australian Institute of Criminology, 2003).
White, R. *Hanging Out – Negotiating Young People's Use of Public Space,* report no. 7 (Canberra: National Crime Prevention Attorney-General's Department, 2000). See: www.ncp.gov.au/ncp/publications/no7/index.htm
White, R. and Coventry, G. *Evaluating Community Safety – a guide* (Melbourne: Darebin City Council and Victorian Department of Justice, 2000). See: www.lawlink.nsw.gov.au
Wilkie, W. (1998) *Treating Post Traumatic Stress Disorders Caused by Bullying,* in McCarthy, P. Sheehan, M. Wilkie, S. and Wilkie, W. (eds) *Bullying: Causes, Costs and Cures,* 1st edition (Brisbane: Beyond Bullying Association, 1998) 15–21.
Williams, C. 'A pain in the neck: passenger abuse, flight attendants and emotional labour', *The Journal of Occupational health and Safety – Australia and New Zealand,* 16(5) (2000) 429–35.
Williams, C. Thorpe, B. and Chapman, C. *Aboriginal Workers and Managers: History, Emotional and Community Labour and Occupational Health and Safety in South Australia,* (Henley Beach, South Australia: Seaview Press, 2003).
Williams, K. 'Social ostracism' in Kowalski, R. (ed.) *Adversive Interpersonal Behaviours* (New York: Plenum Press, 1997) 133–65.
Wilson, E. *Sphinx in the City: Urban Life, the Control of Disorder, and Women* (London: Virago Press, 1991).

Wirth, K. 'Bullying and violence in the workplace', in CCH and Freeholds *Australian Master OHS and Environment Guide 2003* (Australia: CCH, 2002) 495–518.

Wolfgang, M. and Feracutti, F. *The Subculture of Violence: Towards an Integrated Theory in Criminology* (London: Tavistock, 1969).

Workers' Health Centre, *Health and Safety Fact Sheet: Violence at Work* (Sydney: Workers' Health Centre, 1999).

WorkSafe Victoria, *Guidance Note – Prevention of Bullying and Violence at Work* (Melbourne: Victorian WorkCover Authority, February 2003).

WorkSafe Western Australia, *Code of Practice: Workplace Violence* (Perth: WorkSafe Western Australia, 1999).

World Health Organisation, *Violence: A Public Health Priority* (Geneva: World Health Organisation, 1995).

World Health Organisation (WHO), *ICD-10* (Göttingen: Huber, 1992).

Wright, B. and Barling, J. 'The executioners' song: Listening to the downsizers reflect on their experiences', *CJAS* 14(4) (1998) 339–355.

Wyatt, J. and Hare, C. *Work Abuse: How To Recognise And Survive It* (Vermont: Schenkman Books, 1997).

Yamada, D. 'Workplace bullying and the law: Towards a transnational consensus' in Einarsen, S. Hoel, H. Zapf, D. and Cooper, C. (eds) *Bullying and Emotional Abuse in the Workplace: International Perspectives in Research and Practice* (London: Taylor & Francis, 2003) 399–412.

Zadek, S. Sabapathy, J. Dossing, H. and Swift, T. *Responsible Competitiveness: Corporate Responsibility Clusters in Action* (Copenhagan: The Copenhagen Centre, European and Social Affairs DG of the European Commission, January 2003). See: wwww.copenhagencentre.org

Zapf, D. 'European research on bullying at work', in McCarthy, P. Rylance, J. Bennett, R. and Zimmerman, H. (eds) *Bullying: From Backyard To Boardroom,* second Edition (Sydney: Federation Press, 2001) 11–22.

Zapf, D. and Einarsen, S. 'Individual antecedents of bullying: victims and perpetrators', in Einarsen, S. Hoel, H. Zapf, D. and Cooper, C. *Bullying and Emotional Abuse in the Workplace* (London: Taylor and Francis, 2003) 165–84.

Zapf, D. and Gross, C. 'Conflict escalation and workplace bullying: a replication and extension', *European Journal of Work and Organisational Psychology*, 10(4) (2001) 497–523.

Zapf, D. Einarsen, S. Hoel, H. and Vartia, M. 'Empirical findings on bullying in the workplace', in Einarsen, S. Hoel, H. Zapf, D. and Cooper, C. (eds) *Bullying and Emotional Abuse in the Workplace: International Perspectives in Research and Practice* (London: Taylor and Francis, 2003) 103–26.

Zdenkowski, G. Sentencing trends: past, present and prospective', in Chappell, D. and Wilson, P. *Crime and the Criminal Justice System in Australia: 2000 and Beyond* (Sydney: Butterworths, 2000).

Zimbardo, P. *Psychology and Life*, 12th edition (New York: Harper Collins, 1988).

Legal cases cited:

State of New South Wales v Seedsman [2000] NSWCA (12 May 2000).

Index